Problem Soils and Their Management

NEW INDIA PUBLISHING AGENCY
New Delhi – 110 034

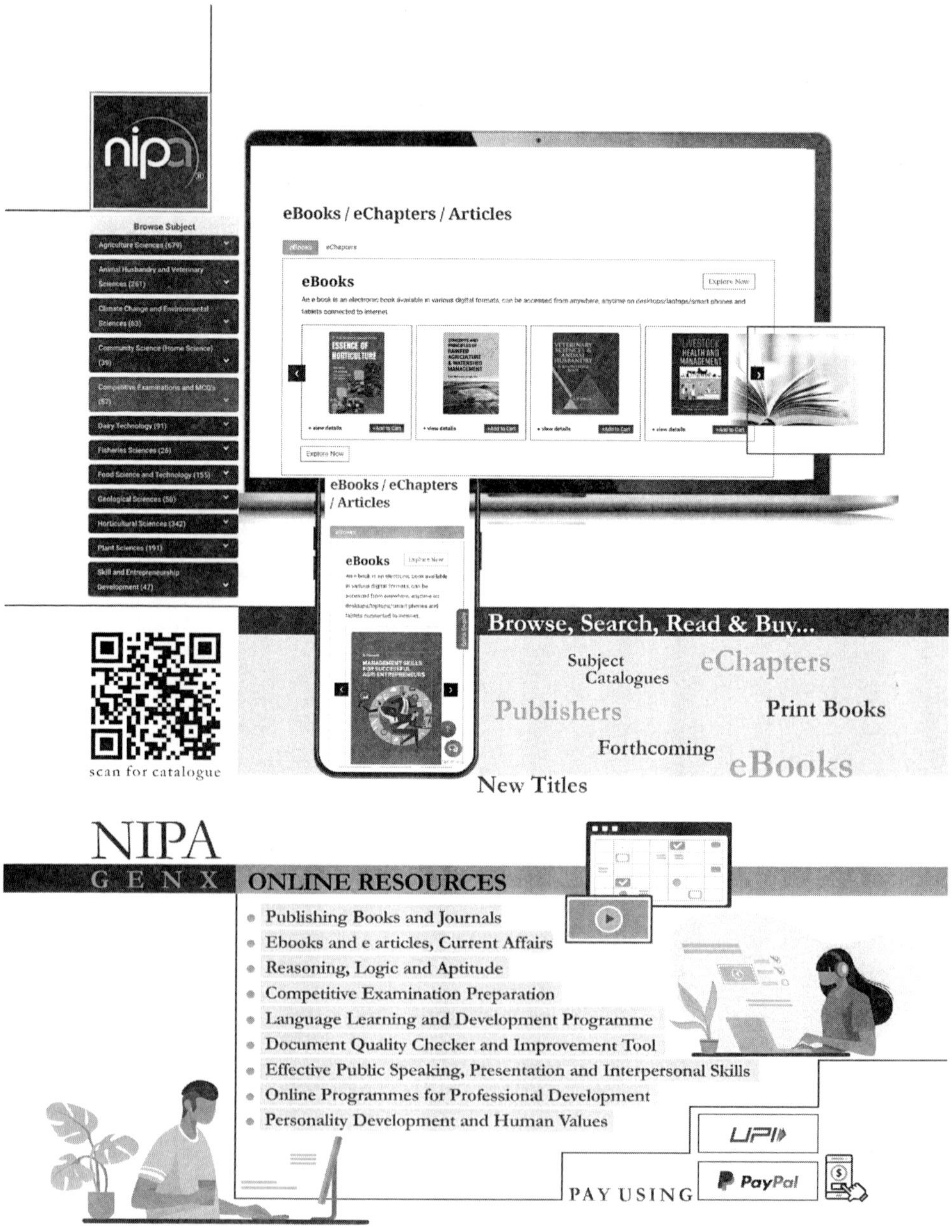
nipa
Browse Subject
Agriculture Sciences (679)
Animal Husbandry and Veterinary Sciences (261)
Climate Change and Environmental Sciences (63)
Community Science (Home Science) (39)
Competitive Examinations and MCQ's (57)
Dairy Technology (91)
Fisheries Sciences (28)
Food Science and Technology (155)
Geological Sciences (50)
Horticultural Sciences (342)
Plant Sciences (191)
Skill and Entrepreneurship Development (47)
eBooks / eChapters / Articles
eBooks
Explore Now
eBooks / eChapters / Articles
scan for catalogue
Browse, Search, Read & Buy...
Subject Catalogues
eChapters
Publishers
Print Books
Forthcoming
eBooks
New Titles
NIPA
GENX
ONLINE RESOURCES
Publishing Books and Journals
Ebooks and e articles, Current Affairs
Reasoning, Logic and Aptitude
Competitive Examination Preparation
Language Learning and Development Programme
Document Quality Checker and Improvement Tool
Effective Public Speaking, Presentation and Interpersonal Skills
Online Programmes for Professional Development
Personality Development and Human Values
PAY USING
UPI
PayPal

Problem Soils and Their Management

Dinabandhu Jena

Professor and Head
Department of Soil Science and Agricultural Chemistry
Institute of Agricultural Sciences
SOA Deemed to be University
Bhubaneswar-751 030, Odisha

and

Formerly Professor and Head
Department of Soil Science & Agricultural Chemistry
College of Agriculture
Orissa University of Agriculture and Technology
Bhubaneswar - 751 030, Odisha

nipa®

NEW INDIA PUBLISHING AGENCY

New Delhi – 110 034

NEW INDIA PUBLISHING AGENCY
101, Vikas Surya Plaza, CU Block, LSC Market
Pitam Pura, New Delhi - 110 034, India
Email: info@nipabooks.com
Web: www.nipabooks.com

For customer assistance, please contact
Phone: + 91-11-27 34 17 17 Fax: + 91-11- 27 34 16 16
E-mail: feedbacks@nipabooks.com

ISBN : 978-81-19476-84-1

Composed and Designed by NIPA.

Preface

This book has been written as a text book for the students to have full grasp of nature and causes of Land Degradation, methods of evaluation, characterization and control measures for maintaining its productivity. The vast area of land degradation has been covered in a simple and concise form. The book offers sufficient materials for the students to prepare entire course of **Problem soils and their management** based on ICAR Syllabus and also for the students of allied agriculture and environmental sciences.

Although much of research papers on the subject have been published, teachers of the subject have been handicapped by lack of a suitable text book. The students also faced problems when they are asked to refer several books for this course. since, none of the books so far published contain all the chapters as per ICAR syllabus.

Having more than forty years of teaching experiences on different courses namely soil physics, soil erosion and conservation, problem soils and their management, soil, water and air pollution, acid soils, land degradation and restoration at under graduate, post graduate and Ph.D levels, I am fortunate to go through several text books, review papers and published articles in different bulletins. Practical experiences in various aspects of the subject and fed back from the students through assignments has resulted in preparation of a complete set of notes which has been used to build up a course on **Problem soils and their management**.

The book comprises of fifteen chapters covering all aspects of physical, chemical and biological processes leading to land degradation under different agro–ecosystems. Physical land degradation leading to soil compression, compaction, soil crusting, slow and highly permeable soils were discussed thoroughly. Chemical land degradation leading to iron and aluminium toxicity, soil acidity, acid sulphate soils, soil salinity and alkalinity have discussed. Equal emphasis has also been given to assess and monitor the extent of land degradation and its impact on soil and water using modern technology–Remote Sensing, GIS and GPS. For management of wastelands, the concept of watershed management based on land capability classification, land irrigability classification and land suitability classification were discussed in detail. Due emphasis was given to maintain soil health and soil quality for sustainable crop production.

At the end of each chapter, study questions have been presented for the benefit of the students.

It is hoped that this book will cater to the needs of the students pursuing agriculture and allied sciences and also the related subjects.

The author will be grateful to the readers giving suggestions for further improvement of the subjects.

I duly acknowledge the sources of the diagram and tables that have been reproduced from text books and other publications.

Author

Contents

1

Wastelands Distribution Formation Management Challenges and Strategies

I. Definition of wasteland

Wasteland include degraded forest, overgrazed pastures, drought- struck pastures, eroded valley, hill slopes, water logged marshy lands, barren lands etc.

Different authors have defined wastelands as:

a) Land which is lying unproductive or which is not being utilized to its potential.

b) Land which is incapable of producing material or services of value (American Society of Soil science).

c) Land which has been abandoned and for which there is no further use (Mine spoils).

d) Land which produces less than 20% of economic potential.

e) Land where no greenery can be sustained.

f) Land which is neither under forest cover or agricultural cover or assigned for specific purposes such as National park or National hydro power projects.

g) The Technical Task Group Report of the National Wasteland Board defines the wasteland due to different constraints.

II. Classification of wastelands

Wastelands are classified under two groups:

i) Culturable wastelands and

ii) Unculturable wastelands

i) Culturable wastelands

These are cultivable wastelands which are not being utilized to their full potential or are being mismanaged due to various reasons such as state or private occupation or having being declared as notified forest area. Such cultivable wastelands include gullied lands, surface water logged marshy lands, undulating and saline land, degraded forestry and pastures, shifting cultivation areas, sand dunes or mining spoils etc.

ii) Unculturable wastelands

These are the wastelands which are not available for cultivation. These includes barren rocky lands, steep slopping areas and areas covered by snow or glaciers.

Significance of wastelands

Formation of wastelands lead to the deterioration of the ecological balance by adversely affecting the various components of the ecosystem directly or indirectly dependent on that particular land.

III. Distribution of wasteland

Wastelands are lands which are economically unproductive, ecologically unsuitable and subject to environmental deterioration. The official estimates of wasteland in 2004 was nearly 63.85 million ha- more than 20% of the country's geographical area.

The category wise wasteland types of India is shown in table 1.1.

Table 1.1. Category wise wastelands of India (2004).

S. No.	Category	Total waste land (Area in Sq. km)	% of total geographical area
1.	Gullies and/ or ravenous land	20553.35	0.65
2.	Land with or without scrub	194014.29	6.13
3.	Water logging and marshy land	16568.45	0.52
4.	Salt affected land	20477.38	0.65
5.	Shifting cultivation	35142.20	1.11
6.	Underutilized / degraded notified forest land	140652.31	4.44
7.	Degraded pasture / grazing land	25978.91	0.82
8.	Degraded land under plantation crop	5828.09	0.18
9.	Sands – Inland / Coastal	50021.65	1.58
10.	Mining / Industrial Wasteland	1252.13	0.04
11.	Barren rocky / stony wasteland / sheet rock area	64584.77	2.04
12.	Steep slopping area	7656.29	0.24
13.	Snow covered and / or glacial area	55788.49	1.76
	Total wasteland area	638518.30	20.17

Source: 1,50,000 scale wasteland maps prepared from landsat Thematic Mapper/IRS LISS 11/111 data

According to state wise break up, the state with the greatest percentage of wasteland is in Rajasthan with 45%, followed by the hill states Jammu and Kashmir, Tripura, Himachal Pradesh, Manipur, Sikkim, Mizoram, Meghalaya, and Nagaland (Table1.2).

Table 1. 2. Aerial estimation of wasteland (statewise) in 2000.

S. No.	State	Wasteland Area (ha)	Percentage of State Area
1.	Rajasthan	154884.538	45.258
2.	Jammu and Kashmir	72695.461	32.71
3.	Tripura	3253.8	31.06
4.	Himachal Pradesh	15044.44	27.02
5.	Manipur	5665.44	25.34
6.	Sikkim	1843.44	25.26
7.	Mizoram	4478.76	21.24
8.	Meghalaya	4746.72	21.11
9.	Nagaland	3253.80	19.69
10.	Gujarat	3713.16	8.40
11.	Arunachal Pradesh	13436.28	16.1
12.	Uttar Pradesh	40768.00	13.85
13.	Kerala	5320.92	13.69
14.	Tamil Nadu	17610.80	13.54
15.	Odisha	19484.51	12.51
16.	Karnataka	21934.44	11.44
17.	Andhra Pradesh	30662.27	11.10
18.	Madhya Pradesh	54137.04	9.97
19.	Maharashtra	30662.28	9.96
20.	Goa, Daman and Diu	344.52	9.03
21.	Haryana	3713.16	8.39
22.	Bihar	15312.0	8.81
23.	Punjab (Including Chandigarh)	3904.5	7.73
24.	West Bengal	6334.48	7.23
25.	Assam	2947.56	3.75

IV. Formation of wasteland

Wastelands are formed either by natural ways or by man-made activities (Table 1.3). The undulating upland, sandy area, snow covered area, coastal saline area etc. are natural formation while, gullied or ravenous land, Jhum or forest blank, barren hill-ridges etc. are formed by man-made activities.

There are four major anthropogenic activities that lead to formation of wasteland areas viz.:

i) Deforestation leading to soil and water erosion

ii) Overgrazing

iii) Over cultivation and

iv) Unskilled irrigation leading to water logging.

In addition, both cultivated and uncultivated lands suffer from soil erosion and degradation. Mining is another important activity that causes deforestation and land degradation too. Desert is one of the major natural wasteland area and any activity relating to desertification is considered to be one of the major activities underlying the wasteland formation.

Table 1. 3. Categories of wasteland based on causative agents.

Water	Wind	Man	Others
Splash erosion	Sand dunes	Mine spoils	Land slides
sheet erosion	Sand bar	Shifting Cultivation	Shallow soil
Rill erosion	Coastal sand	Industrial wasteland	Stony soils
Gulley erosion			
Ravenous land			
Saline soil			
Marshy land			
Water logged			
Alkali soil			

Various agencies have been engaged in inventorizing and characterizing soil resources for developing plans at National, State and regional levels which are presented in table 1.4.

Table 1.4. Estimation of soil degradation in India by various agencies.

S. No.	Agencies	Estimated extent (Mha)
1.	National Commission of Agriculture (NCA, 1976)	148.09
2.	Ministry of Agriculture (1978) (Soil and water conservation Division)	175.00
3.	Society for Promotion of Wasteland Development (SPWD, 1984)	129.58
4.	NRSA (1985)	53.28
5.	Ministry of Agriculture (MOA, 1985)	173.64
6.	Ministry of Agriculture (MOA, 1994)	107.43
7.	NBSS & LUP (1994)	187.70

The NBSS & LUP, Nagpur has furnished latest data on soil degradation on 1:250,000 scale under different category. This is a further refinement on the existing mapping on 1:4.4 million scale under Global Assessment of Soil Degradation (GLASOD) guidelines. The indicators of degradation were water erosion, wind erosion, water logging / flooding, salinity / alkalinity, acidity etc. As per the estimates, about 45% of total geographical area of the country is degraded due to different degrading agents (Table 1.5).

Table 1.5. Categories of land degradation problem in India.

S. No.	Type of degradation	Total degraded area (000, ha)
1.	Water erosion	93680
2.	Wind erosion	9483
3.	Water logging	14299
4.	Salinity/alkalinity	5944
5.	Soil acidity	16033
6.	Complex problem	7381

Source: NBSS & LUP (2004)

The Ministry of Agriculture, Soil and Water Conservation Division has estimated severe eroded area in the country. According to their estimate, out of the total geographical area of 328 million ha, about 175m ha (about 53% of total area) has been affected due to various reasons such as water and wind erosion, salt affected soils, water logging etc. (Table 1.6). Out of this area, 150 m ha is eroded by water and wind and 25 m ha by other mechanisms.

Table 1.6. Categories of degraded land in India.

S. No.	Category	Area (m ha)
1.	Rainfed non paddy areas	82.11
2.	Other cultivable land (including permanent pastures and cultivable wasteland)	22.56
3.	Forest land (protected and unclassified forests)	19.49
4.	Area not available for cultivation (nonagricultural use of wasteland)	7.91
5.	Gullies and ravines	3.97
6.	Saline and alkaline	6.97
7.	Shifting cultivation	2.85
8.	Ravine lands (khadar / char / bill)	2.48
9.	Torrent and stream banks	0.25
10.	Desert	18.79
11.	Coastal sandy land	1.63
12.	Water logged areas	5.99
	Total	175.00

V. Prospects of wasteland development in India

According to the National Wasteland Development Board, wasteland mean degraded land which can be brought under vegetative cover with reasonable effort and which is currently lying unutilized and land which is deteriorating for lack of appropriate water and soil management or on account of natural cause.

Integrated Wasteland Development Project (IWDP) was launched in India during 1989-90 with an objective for the integrated wasteland development based on village/micro watershed plans. These plans are prepared based on the land capability, site condition and local needs of the people. The major activities of the scheme were:

i) In situ soil and moisture conservation.

ii) Planting and sowing of MPT, shrubs, grasses, legumes and pasture land development.

iii) Encouraging natural regeneration.

iv) Promotion of agroforestry and horticulture.

v) Wood substitution and fuel wood conservation measures.

vi) Awareness raising, training and extension.

vii) Encouraging people's participation.

viii) Drainage line treatments.

ix) Afforestation of degraded forest and non-forest wasteland.

x) Development and conservation of common property resources.

Reclamation of wasteland

The reclamation and development of wasteland has four major ecological objectives:

i) To improve the physical structure and quality of the soil

ii) To improve the availability and the quality of the soil

iii) To prevent the shifting of soil, land slides and flooding and

iv) To conserve the biological resources of the land, land slides and flooding.

There are a number of ways by which wasteland can be reclaimed for productive purpose. Among them (1) Wasteland afforestation (2) Pasture development and (3) Watershed management programme are noteworthy.

1. Afforestation in wasteland

Besides the soil and water conservation measures, various types of plantation programme including energy plantations, fuel and fodder spp. and dry land fruit crops can be taken up on the wasteland areas depending on soil and climatic conditions.

The plantation programmes as recommended by the Social Forestry Department can be taken up with people's participation.

The programmes are

a) Plantation on Government wasteland / Community lands

b) Plantation on degraded barren hills and erosion – prone areas

c) Strip plantations

d) Reforestation and rehabilitation

e) Farm forestry and

f) Plantation of dry land horticultural fruit trees.

a) Plantation on Government wasteland/Community lands

With the technical support of Govt. Organizations, the villagers will carry out tree planting on community lands in the vicinity of villages. These areas are to be planted with fuel wood, food, fodder, fruit, minor forest produce and small timber species. Grasses and fodder crops can be taken up as inter crops to meet the pressure on grazing. The villagers are to manage the plantation programme by themselves with full protection against grazing.

b) Plantation on degraded barren hills and erosion prone areas

A large number of tribal population depend on shifting cultivation, locally known as PODU, for their living. Many villages are also surrounded by degraded barren hills, where depletion and deterioration of vegetation has created a serious situation. The plantation programme will be taken up to achieve:

i) Arrest degradation of soil fertility of such hill slopes.

ii) Undertake intensive soil conservation measures to conserve soil and moisture which will be beneficial for tree growth.

iii) Provide vegetative cover to the barren hill slopes and other erosion prone areas.

Tree spp. will be planted in staggered contour trenches of size 2.5 m length x 0.5 m depth x 0.5 m width or in pits of size 0.45 m x 0.45 m x 0.45 m. Suitable tree spp. will be planted in pits and trenches. Planting of trenches will be supplemented by sowing of grasses, legumes and fodders and other soil binding species. Gully plugging and nalabunding will be done in affected areas.

c) Strip plantations

Strip plantation programme will be undertaken on government lands adjoining village and Panchayat road ways, canals, tanks etc. The primary objectives of this programme are:

i) Increase the supply of fuel, fodder, fruit and other forest products for local community,

ii) The poor villagers collect twigs, branches and fallen dry leaves

iii) Provide employment for poor villagers and

iv) Provide environmental benefits in form of shade etc.

Thorny species like acacia, ber, prosopis and similar species which do not require management care will be preferred for planting.

d) Reforestation and rehabilitation

A large areas of forest in the village vicinity have been reduced to a bushy stage because the villagers used these forest trees as fuel wood. These areas will be reforested and rehabilitated through the villagers to generate production of high value timber and forest produce. The existing plant species will be retained with complementary planting of high value tree species like sal, piasala, teak etc.

e) Farm forestry

Farm forestry would aim at assisting individual farmers and/or institutions (School, College, Club etc.) to plant fuel wood, small timber, fodder and fruit trees. Prime agricultural land will not be diverted to farm forestry plantations.

Tree planting raised on farmers land not suitable for agriculture, on homesteads and field bunds can increase farmers income. Farm forestry including fodder, fruit, wood and other products has the potential for meeting the objectives of improving the standard of the large majority of small land holders.

f) Dry land horticultural fruit trees

Dry land horticulture has great potential in rainfed wastelands. Successful cultivation with higher economic returns from fruits and vegetables in arid and semiarid areas need special technological measures viz. nursery raising, establishment, soil and moisture conservation, postharvest technology and marketing. There is tremendous scope of growing hardy fruits like Ber, Amla, Pomegranate, Custard apple, Jamun, Fig, Datepalm, Guava etc. in rainfed as well as with supplementary irrigation.

The desirable characteristics of fruit trees in these areas should have

i) Deep and extensive root system (Ber, Mango),

ii) The characters like thick cuticle, hairy leaves, wax coating on leaf surface and covered stomata (Fig, Ber),

iii) Flowering and fruiting should synchronizes with maximum availability of soil moisture and

iv) Tolerance to salinity and alkalinity (Date palm, Ber, Amla).

2. Pasture Development

Importance of pasture grasses is three fold i.e. livestock feed, soil builder and aid in soil conservation.

Fibrous root system of grasses bind the soil particles and cover the land surface by their spreading habit and this prevent soil erosion by wind and water.

The suitable grasses and legumes for different agro-climatic zones are given below:

i) Rainfall < 200 mm with sandy soil-*Panicum antidotale, Lasiurus sindicus etc.*

ii) Rainfall 200-400 mm with alluvial soils-*Cenchrus ciliaris, Clitoria ternatea etc.*

iii) Rainfall 400-600 mm with red loam soils-*Cenchrus ciliaris, Lablab Purposes etc.*

iv) Rainfall more than 600 mm with loam to clay loam soils-*Panicum maximum, Napier hybrid* etc.

The yield of fodder crops can be increased by adopting better techniques viz., selection of good variety, suitable for multiple or mixed cropping, use of fertilizer and water management practices.

3. Watershed-Concept and Management

Watershed is a geo-hydrological unit or a piece of land that drains at a common point. A watershed comprises of arable lands, non-arable lands and natural drainage lines in rainfed areas. A watershed may be a large area having a number of villages, roads, hills, fields and forests within it, the salient feature being the drainage of all precipitation through one outlet. The watershed development programme is not confined to agricultural land alone, but covers the area, starting from the highest point of the area (hill top) to the outlet of the natural stream. A watershed development programme focus on different problems related to soil, vegetation, water and animal. The process of setting regional priorities for watershed development programme should be based on careful consideration of:

a) Physical evidence of degradation (extent of gully, sheet erosion, run off rates, deforestation and water quality).

b) Socio economic indicators of degradation (fodder and fuel efficiency, decrease in farm yield).

c) Livelihood of introducing sustainable technical packages (slope, soil characteristics, current land use practices, localwillingness).

The approach and strategy of a watershed development programme based on two concepts such as:

A. Integrated watershed development and

B. Sustainable farming system

A. Integrated watershed development

Spatially watershed consists of three physical sectors:

i) Arable or cultivated land which are privately owned,

ii) Non-arable lands which includes pasture, grazing lands and barren lands and

iii) Network of natural drainage lines.

These three sub-sectors would be treated as one geo-hydrological unit for project planning and implementation to ensure sustainable use of natural resources of land and water.

B. Sustainable farming system

A typical rainfed farmer derives his income from his own cultivated land and partly from community land in addition to the income from livestock. In some areas, specialized activities like sericulture, bee keeping, mushroom cultivation, lac cultivation, fishery and collection of forest products also supplement the income of farmers.

Thus the project will treat the following sub-components of the household farming systems:

i) Food sub component

ii) Fodder sub component

iii) Fuel sub component and

iv) Income generation component

The inter–relationship among various components and sub-components of the farming system will be treated as a part of one organic unit.

Community farming systems

There are many community owned resources like common village ponds for fishery and irrigation, village wood-lots etc. The project would analyze and treat the inter-dependence of individual farming systems with community production system.

People's participation

Participation of villagers in project planning, project preparation, implementation and post project evaluation would be an integral part of the approach and strategy.

The watershed development programme is multidisciplinary in nature. This usually requires close interaction and co-ordination among different line department viz. forestry, horticulture, agriculture, animal husbandry, soil and water conservation and revenue.

Soil and water conservation measures

The project would rely on low cost and vegetative conservation measures in place of engineering structures which are costly to install and difficult to maintain.

The objectives of soil and water conservation measures are:

i) To minimize or control surface run off and soil erosion from various lands

ii) To ensure adequate moisture supply for biomass production either by increasing infiltration rate in soil and / or manipulation of runoff water

iii) To store excess run off water locally for subsequent use

Soil and water conservation measures for agricultural lands

i) **Slopes less than 3%**-Contour cultivation, contour farming with suitable crop geometry and cultural practices.

ii) **Slopes 3-6%**-Contour bunding where soils have good infiltration and permeability rates with less than 500 mm annual rainfall. Graded bunds maybe adapted on soils with low infiltration rates and rainfall exceeding 500 mm annually.

iii) **Slopes between 6-33%** - Where the soils are favourable, contour trenching, terracing can be adopted. For raising crops like potato, bench terracing with suitable vertical intervals may be adopted on steep sloppy lands.

Soil and water conservation measures for nonagricultural lands

These lands are unsuitable for agricultural crops because of greater slope, erosion, rockiness, shallowness of soils. These lands can be developed into suitable fodder -fuel-fibre-fruit-timber areas which will make them remunerative.

Control of ravines

Ravine areas occupy nearly 4.0 million ha in our country. These can be reclaimed by constructing composite earthen and brick masonry check dams in the beds and bench terracing on sides. In these reclaimed lands, excellent orchard could be raised.

Run off harvesting, storage and recycling

Harvesting of runoff water at a micro level for storage and recycling is necessary for better utilization of rainfall. This will reduce loss of water and soil.

Proper land use in micro watershed

Proper land use in micro watershed include:

i) Soil conservation measures

ii) Cropping pattern based on soil characteristics and availability of irrigation including agro forestry

iii) Water harvesting measures

iv) Improvement of soil fertility

v) Other suitable measures includes silviculture, silvipastoral, horticulture, agro forestry etc. depending upon the soil characteristics.

Some of the measures are

i) Soil conservation measures for crop land

Bunding of field, mulching, growing grasses and legumes and agronomic practices.

ii) Soil conservation measures for wasteland

The measures include

- Provide vegetal cover at the earliest
- Bunding
- Gully plugging
- Contour trenching / stone dykes / vegetative barriers.

Challenges and issues

Challenges

The management of our natural resources is essential for livelihood, environmental and socio economic security of the country. The decline in total and partial factor productivity of agriculture due to land degradation, rising and falling of water table, secondary salinization and pollution of soils etc. are major challenges in Indian Agriculture. The decline in size of land holding from 0.48 ha per capita in 1950 is likely to become 0.10 ha in 2050 because of the increasing population pressure from 350 to 1500 million in 2050 is another of deep concern.

India has no doubt made a spectacular achievement by increasing the food grain production from 51 Mt. in 1950 to 283.37 Mt. in 2018-19 from the net cropped area of about 141-142 Mha but, remained stagnated due to various reasons. But there is scope to convert wasteland (120 M ha) into cultivatable land. The past achievements in attaining self-sufficiency in food grain production has jeopardized the abilities to meet future challenge in the present scenario. There is hardly any scope to increase the area under cultivation as it has already been extended to marginal soil. On the other hand we are continuously losing agricultural land to non-agricultural used in addition to erosion and degradation.

Deforestation followed by intensive cropping and grazing resulted in accelerated erosion of soils with subsequent loss of soil and nutrients out of ecosystem. Problems of land degradation are seriously affecting our ability to increase and sustain agricultural productivity.

The above challenges and concern demand, concentrated efforts to minimize land degradation and simultaneously maintain the ecological balance and soil quality for sustainable agricultural production.

Strategies

a) Assessing and monitoring

For judicious planning and development, we need proper database which is often not available. Modern technologies like Remote Sensing, GIS, GPS should be harnessed for creating desired database which should be accessible, utilizable and used in day-to-day functioning right from the grassroots level upto the laboratory. These tools would help us in monitoring the progress made in land and soil reclamation and restoration. For achieving this goal, the following issues need implementation as measures for long-term planning:

- Maintaining eco balance by arresting deforestation and promoting area under natural forest to control pollution and restore ecological balance.
- Maintaining soil resources, especially of land degradation problem and status.
- Developing standards and methodology for uniformity in soil mapping to meet users need.
- Digitization of maps, efficient storage of data base and timely delivery to the land users.
- Assessing risk of land degradation to take preventive measures.

A short term goal, there appears an urgent need to take out 20-25 Mha of currently marginal land under cultivation and bring it under permanent vegetative cover.

b) Watershed based approach

There is a need to adopt an integrated watershed based approach to the management and development of natural resources including soil, water, crop, range forests and livestock in order to:

- Conserve and upgrade crop lands and manage degraded lands.
- Develop and demonstrate site specific technology for soil and moisture conservation towards sustainable crop production.
- Augment the fodder, fuel and fruit resources through alternate land use systems.

c) National Land use Policy (NLUP)

The main objectives of a National Land Use Policy are:

i) To meet the consumption needs of the growing population by increasing productivity of integrated land resources in the country

ii) To coordinate formulation and implementation of water resources management, forest management and urban planning policies.

iii) To prevent soil degradation and to restore the productivity of the degraded lands.

In view of the above objectives the successful implementation of NLUP requires to

- Prepare a character of soil resources of the country to know the kind and extent of different types of soils.
- Prepare thematic maps of each set of land to know the potential and constraints.
- Develop a soil site suitability model for different crops for suggesting alternative land uses.
- Plan soil conservation/amelioration techniques for such areas needing treatment for optimizing land use.
- Testing and transfer of agro-technique through front line demonstrations.

Study Questions

1. Define wasteland. What is difference between culturable and unculturable wasteland?
2. Name different types of wastelands based on causative agents.
3. Explain how anthropogenic activities leading to wasteland formation is more serious than natural ways?
4. Why higher percentage of wasteland are seen in Rajasthan and North hill states of India to rest of the country- give reasons?
5. Name four ecological objectives of wasteland development in India.
6. What are the beneficial effects of pasture grasses in reclamation of wasteland?
7. Define watershed. Name different physical components of a watershed.
8. Name different sub components of a sustainable farming system and their significance.
9. What are the basic objectives of soil and water conservation measures? State some conservation measures for agricultural land having 3-6% slope.
10. Briefly discuss the beneficial effects of legumes and grasses to control soil and water erosion.
11. Briefly discuss- the success of watershed management depends upon people's participation.
12. State the objectives and features of a National land use policy.
13. The net cultivated area in India varied between 140-142M ha. Is there any scope to increase the area under cultivation?

2

Soil Physical Constraints Leading to Land Degradation

Crusting soils

What is soil crusting?

Soil crusting refers to formation of a compact thin layer having high bulk density and high penetration resistance at the soil surface when dries after dispersion of fine particles and clogging of pores due to beating action of rain drop and slaking of irrigation water.

In arid and semiarid regions, crust formation occurs due to the destruction of soil aggregates. Crust formation occurs in all types of soils except sand. It is characterized by high bulk density, low non capillary pores, low hydraulic conductivity and aeration and high penetration resistance.

Important soil properties that encourage the crust formation

i) Low organic matter content

ii) High salt content

iii) High exchangeable sodium content

iv) High content of oxides of iron and aluminium

Mechanisms of crust formation

Soil crust develops as a result of compaction on the soil surface due to an externally applied force.This force is supplied by beating action of rain drops and the radiant energy of sun. Due to beating action of rain drops on dry soils, there is slaking of the soil aggregates followed by dispersion and orientation of the finer clay particles. The clay particles mixed in rain water, make water muddy, percolated down ward, swells under wet condition and clog the soil pores. The muddy water on the soil surface mixed with iron and aluminum forms an impervious compact zone of higher bulk density when the soil dries.

The beating action of raindrop on soil surface is related to its kinetic energy- Ek:

Ek = ½ m v^2 where, m is the mass of rain drop and v is drop velocity.

The total action of a storm is a function of the kinetic energy of all the rain drops:

Ek Total = ½ $\sum \mathbf{m_i v_i^2}$

Where, $\mathbf{m_i}$ **and** $\mathbf{v_i}$ are the masses and velocities of rain drops during a storm.

The falling of the rain drops due to gravity is encounters by the viscous resistance of the atmosphere. When resistance is equal to the downward force of gravity, the rain drop falls at a constant velocity called **terminal velocity** which depends on mass. The impact of rain drops varies with size of the drop.

Another factor affecting crust formation is the sodium content in soil. A high sodium absorption ratio (SAR) encourages dispersion and swelling of clay particles on surface layer. The migrated clay particles clog the pores beneath the surface. Extreme clay dispersion and clogging can occur at SAR values as low as 10. Surface crust greatly reduces the water intake by soil even if the crust is thin (more than 1mm). The permeability of surface soil is reduced by 2000 times than the undisturbed soil below. The zone below the surface consists of migrated clay particles with very isolated air pores.

Seedling emergence and soil crusting

The effect of soil crusting on seedling emergence depends on crust thickness and strength, as well as the size of the seeds and vigour of the seedlings. Small size seeds are more susceptible to crusting. With increasing the crust strength, the seedlings buckle, grow in a horizontal direction and fail to emerge. The seedlings of grain sorghum decreased if the soil strength exceeds 3 bars as measured with a penetrometer and ceased when the strength reached 13 to 18 bars.

Strong crust with high bulk density restricts aeration under moist condition and prevents effective diffusion of oxygen into soil. Lack of aeration became a problem in germination of the seeds.

Evaluation of crust strength

The strength of soil crust can be evaluated by modulus of rupture test and resistance to penetration of a soil profile. These tests were designed to imitate the process by which a seedling forces its way upward by penetrating and rupturing the crust. Using modulus of rupture test, the emergence of bean seedlings in a fine sandy loam was reduced from 100 to 0 per cent, when the crust strength increased from 108 to 273 mbar. The emergence of sweet corn was prevented when the crust strength exceeded 1200 mbar. However, the crust strength which prevents emergence depends on crust thickness, soil wetness as well as the size of seeds and depth of placement.

Control Measures

Soil crusting can be controlled by surface mulches which prevents the beating action of rain drops. Use of lime ($CaCO_3$), gypsum, phosphatic fertilizers reduce cementing action of iron and aluminium in red and laterite soil and reduce crust strength. Use of bigger size seeds increases per cent of emergence. Addition of organic matter which promotes stable aggregate formation and resist dispersion, prevent crust formation.

Soil Compression, Puddling and Compaction

Compression

Compression is defined as the change in volume of a soil under an applied force. The volume of soil consists of the solid phase and the voids between it. Under an applied pressure, the void ratio decreased. The relationship between void ratio (e) and applied pressure (P) is given by:

$e = A \log P - C$ where, e is the void ratio, A is the compression index (de/ dlog P), P is the applied pressure or load and C is a constant (equivalent to void ratio per unit load, P).

Soil compression occurs due to particle orientation and changes in size of the colloidal micelles. Under dry condition, particles oriented and come closer with an applied pressure. The void ratio in a compressed soil decreased linearly with increasing the pressure. In case of wet-compacted soil, the spacing of the oriented particles are decreased with increasing pressures. The decrease in void ratio in wet compacted soil is higher than the dry-compacted soil at an unit applied pressure.

Compression increases with moisture content to a maximum and then decreases as the amount of moisture become larger. The maximum compression occurs at plastic range of moisture content. The relationship between the compression and pressure is given by:

$Y = a e^{-bx}$ where, Y is the amount of compression, x is the applied pressure, a and b are constants.

Soil puddling

Puddling is the reduction in apparent specific volume (void ratio) of a soil caused by mechanical work done on the soil. The term **puddlability** refers to the susceptibility of soils to puddling. Puddling is done by mixing soil with water to render if impervious. Intensive tillage by repeated ploughing of a wet soil breaks down coarse aggregates and mean particle size decreases. It was reported that after puddling of a lateritic sandy clay loam soil, about 40% of the aggregates break down completely into smaller fractions of 0.005 mm. The mean weight diameter of water stable aggregates decreased from 1.7 mm to 0.36 mm.

The effect of compression and shearing forces on soil puddling is related to moisture content. The effect of shear becomes more prominent as the moisture content increases.

During puddling, the particles become oriented, the non- capillary pores are destroyed and there is reduction in apparent specific volume (void ratio) of soil. The reduction in apparent specific volume dv is used as a measure of Puddlability (P).

The change in apparent specific volume per unit work done (dw) in causing such a change measures puddlability (P) as:

$P = -dv / dw$

The change in volume per unit work done is related to the air filled pore space.

In a puddled soil, the soil structure has been destroyed, the aggregates lost their identity and converted into a structurally homogeneous mass of fine aggregates and textural separates. The degree of disintegration of aggregates depends upon the nature and amount of cementing agents.

The degree of puddling depends on the cultural practices and the nature of soil. High clay content facilitates puddling and high organic matter content increases soil resistance to puddling. Kaolinite clays are more difficult to puddle than calcium saturated clays.

Effect of Puddling on Soil Physical Properties

i) During puddling, coarser aggregates are broken down, non-capillary pore space is destroyed, microvoids increases and apparent specific volume decreases.

ii) Puddling affects the soil water relationship. At a given moisture tension, the moisture content is higher than the normal soil, because puddling reduces the non-capillary pore space.

iii) Puddling decreases the hydraulic conductivity of soil. The hydraulic conductivity of a puddled clay loam soil decreased from 0.192 to 0.054 cm h^{-1}.

iv) Because of decrease in hydraulic conductivity, drainage from puddled soil may be 1000 times slower than that in normal soil.

v) Evaporation from a puddle soil is low.

vi) Puddling creates an anaerobic environment. It increases pH and decreases redox potential (Eh).

vii) Puddling increases water use efficiency and nutrient use efficiency.

Soil Compaction

Soil compaction refers to increase in bulk density of soil due to an applied load or pressure. Soil compaction is a serious and unnecessary form of land degradation leads to soil erosion and decreased crop production

Compaction of soil is the compression of soil particles into a small volume which reduces the size of pore space available for air and water. During compaction, soil particles moves to a closer state of contact and bulk density increases. On loading, the soil solids are rearranged and liquid and gas within the soil pores are compressed.

The state of compaction is defined by the parameters like bulk density, porosity and void ratio. The reorientation of soil particles are given by the relationship:

$e = e_o - C \log P/Po$ where,

e – void ratio, e_o- void ratio at Po and P_0 – initial pressure, P – applied pressure.

The relationship between change in void volume and applied load is exponential, but not linear.

The force required to compact a soil to a given density decreases exponentially with the moisture content. The density of soil at a given moisture content increases exponentially with the force applied. Both these effects are related to the orientation of soil particles.

At a constant pressure, the bulk density increases with increasing water content, then reaches a peak called **maximum density** and wetness value called **optimum moisture** content and there after the density decreases.

Compaction Problem in Agriculture

The use of agricultural machinery for plowing, puddling of land and harvesting of crops exert pressure or load on soil surface. The exert pressure is distributed over a sizable area forming an arch like pattern. The arch action is the tendency of soil to vector out a comprehensive force developed due to friction of soil on soil, interlocking of the particles and the cohesion of water films.

The pressure distribution in soil depends on

i) The amount of load

ii) The size of the contact area between tire and soil

iii) The distribution of pressure within the contact area and

iv) The moisture content and density.

Sole formation

A compacted layer at the bottom of the zone of plowing is called **plow sole**. Plow sole develops due to tillage action and other farm operations. Since the zone above this layer is plowed regularly, the compact layer is observed below the loosened layer. It was observed that all kinds of tillage implements created a sole of some kind under moist condition termed as plow sole, subsoil soles, disk -harrow soles and traffic soles. All these soles reduced the permeability of soil to water and restricted root penetration and elongation.

Effects of Soil Compaction on Soil and Crops

Soil compaction affects soil physical, chemical and biological properties, as well as crop growth and soil health.

i) Soil Bulk Density

The primary effect of soil compaction is an increase in the bulk density of soil. Whenever the bulk density exceeds a certain level, root growth is restricted. The optimum bulk density for root growth is presented in table 2.2.3.1.

Table 2.2.3.1. Ideal and root – restricting bulk densities.

Soil texture	Ideal bulk density (g cm^{-3})	Bulk density restricting root growth (g cm^{-3})
Sand, loamy sand	< 1.60	> 1.80
Sandy loam, loam	< 1.40	> 1.80
Sandy clay loam, clay loam	< 1.40	> 1.75
Silt, Silt loam	< 1.30	> 1.75
Silt clay loam	< 1.40	> 1.65
Sandy clay, silt clay	< 1.10	> 1.58
Clay	< 1.10	> 1.47

Source: USDA (1999)

No till soils often have a high bulk density than recently tilled soils. However, because of higher organic matter content in the top soil and greater biological activity, the structure of a no– till soil may be more favourable for root growth than that of a cultivated soil, despite of higher bulk density.

ii) Porosity

Increase in bulk density decreases soil porosity. Macropores (large pores) required for water and air movement in soil are primarily affected by soil compaction. It is reported that plant roots require about 10 percent of air filled porosity for better growth. The reduction in air filled porosity due to compaction negatively affects root growth and function.

Root penetration is limited if roots encounter much resistance. Penetration resistance is a better indicator of the soil compaction effects on root.

iii) Soil Structure

Soil structure refers to arrangement of soil particles in a definite form or pattern. Soil compaction destroys soil structure and leads to a more massive soil structure with fewer natural voids. In prairie and chestnut soil, granular and crumb soil structure develops due to high organic matter and fine root systems of grasses. Even if when exposed to rainfall, such a soil structure will not break down because the aggregates are very stable and infiltration rate is high. In irrigated command area with rice-rice cropping system, the soil aggregation breaks down due to puddling and the particles reoriented forming a compact layer.

iv) Infiltration and percolation

Infiltration refers to entry of water into soil at soil-atmosphere interface. Percolation refers to the vertical movement of water beyond the rootzone to the water table while seepage is the lateral movement of surface water.

Soil compaction reduces macroporosity resulting much lower water infiltration into soil, as well as a decrease in saturated hydraulic conductivity. However, sometimes unsaturated hydraulic conductivity increases due to soil compaction. Unsaturated hydraulic conductivity is important when water has to move to roots.

If a soil is plowed after compaction, the infiltration rate increases because the soil is cloddy and rough.

v) Soil compaction increases the potential for surface water ponding, water run-off, surface- soil water logging and soil erosion.

vi) Reduces crop emergence as a result of soil crusting.

vii) Impedes root growth and limits the volume of soil exposed by roots.

viii) Reduced crop yield potential.

Management

1. Direct seeding-Direct seeding will eliminate the need for soil cultivation which is the primary cause for development of a hard pan or plow pan layer.
2. Minimize or eliminate soil tillage to prevent soil aggregate break down.
3. Reduce the wheel traffic load on the soil which can be done by keeping axel loads to a minimum.
4. Crop rotation- Avoid rice-rice cropping system. Follow crop rotations including forage, cereals, oilseed and pulse crops that vary rooting depth and types which will reduce soil compaction, increase soil organic matter, promote biological activity, improve water infiltration and penetration into soil. A biologically healthy soil will be more resistant to soil compaction.
5. Ploughing the field with a chisel plough helps to break down sub soil hard pan upto 45 cm depth. This practice should be followed at every 3-5 years.

Slow permeable soils

Slow permeable soils are those having infiltration rate less than 6 cm day^{-1} due to high clay content of soils. In slow permeable soils the rate of water and air movement decreases. The permeability of soil increases with increasing the number of macro pores.

The soil permeability is basically dependent on the pore size distribution in soil. The pore size distribution is influenced by the extent of aggregation in soils. The larger the aggregates, greater is the amount of non-capillary pores. With increase in the size of aggregation from 0.5mm to 5 mm, non- capillary porosity increases from 2.7 to 38.7 percent (Table 1).

Table 1. Relationship of soil porosity to the size of aggregates.

Size of aggregates (diameter in mm)	Porosity in per cent		
	Total porosity	Non-capillary porosity	Capillary porosity
< 0.5	47.5	2.7	44.8
0.5-1.0	50.0	24.5	25.5
1.0-2.0	54.7	29.6	25.1
2.0-3.0	59.6	35.1	24.5
3.0-5.0	62.6	38.7	23.9

Source: Baver (1959)

Soil permeability decreases with depth due to low non-capillary pores because of soil compactness. Low organic matter content in sub-soil decreases soil aggregation. The colloidal properties of clay becomes more dominant in determining permeability in sub-soil.

Permeability increases with coarseness of soil texture. Permeability decreases with increasing fineness of texture, however the extent of aggregation in fine textured soils may override the effects of texture.

The permeability of soil is limited by the presence of plough pans, natural claypans, hardpans or other obstructing layers in a soil profile. The permeability classes for saturated soils are given in table 2.

Table 2. Permeability classes for saturated soils.

S.No.	Permeability class	Inches per hour
1.	Very slow	< 0.05
2.	Slow	0.05-0.20
3.	Moderately slow	0.20-0.80
4.	Moderate	0.80-2.50
5.	Moderately rapid	2.50-5.0
6.	Rapid	5.0-10.0
7.	Very rapid	>10.0

Regur Soils

Slow permeable and poorly aerated black clay is called '**regur**' soils. They are formed in situ by weathering of sedimentary and metamorphic rocks, and have a soil depth varying from 15cm to 240 cm and clay content 30 to 60 per cent. These soils become very sticky and plastic in rainy season and develop wide cracks in summer season because of presence of montmorillonite and bidellite clay. Therefore the workability of the soil is often limited to very short periods of medium water status. Mechanical tillage in wet season causes serious soils compaction. Due to heavy clay content, the capillary porosity is relatively high resulting in impeded drainage and water logging in rainy season. Loss of soil and plant nutrients through run-off water is a major concern in slow permeable soils.

Management of slow permeable soils

The constraints in slow permeable soils can be managed by adopting suitable management practices like:

i) Provision of drainage facilities either through open or closed sub surface drains

ii) Forming contour or compartmental bunding to increase the infiltration rates of soils

iii) Application of organic manures, compost, green manure, composted coir pith, pressmud, gypsum increase permeability in soils.

iv) Adopt ridge and furrow, raised bed or broad bed furrow system.

v) Avoid continuous tillage, since it breaks down soil structure and reduces permeability.

vi) Introducing deep rooted crops, grasses in a cropping system increases permeability.

Highly permeable soils

The highly permeable and low water retentive sandy soils containing more than 70 per cent sand fractions occur in coastal areas, river delta and in desert belts of arid region. The term sandy soils refers to soils whose textural classes are sand, loamy sand or sandy loam. These soils are developed on sand stones, silt stones, unconsolidated sand deposits and recent coastal sands.

Sandy soils are generally very deep, excessively drained and are characterized by porous, massive to structureless structure, high bulk density values. Excessive permeability of sandy soils results in poor water retention capacity, very high hydraulic conductivity and infiltration rates. These soils are devoid of finer particles and organic matter. Very poor soil structure with high non-capillary pores lead to high percolation. The high saturated permeability and low water and nutrient retentivitiy lead to high losses of water and nutrients. The soil pH varies from soil to soil but usually high where the sandy soils are saline,sodic or calcareous. The cation exchange capacity is low and most of the soils are deficient in basic plant nutrients.

Management

The productive capacity of coarse textured soils is limited due to poor water storage and high infiltration and hydraulic conductivity leading to leaching losses of nutrients. Management of these soils for reducing the loss of water and nutrients can be achieved by adopting following technology –

1. The coastal sandy soils are utilized for the growing of cashew nuts, coconuts, bananas, cassava, papaya, guava etc. Application of pond sediments, bentonite clay, compost, sugar factory slurry, compost coir pith, sewage sludge in pits were found effective in reducing percolation losses, enhancing soil water storage and increased plant growth and yield.
2. Sub-soil compaction can be achieved by rolling with a stone roller (400 kg) for 10 to 15 times, 24 hours after irrigation so that soil gets compacted to a desired bulk density.

This technology has been found to be very effective in increasing the nutrient and water retention and increase the yield of crops like bajra, Jowar, arhar, maize, groundnuts, barley and wheat. Instead of stone rollers, drums filled with 400 kg of soil can be rolled.

3. In arid sandy soils, establishment of grape, date palm, sweet orange ber, pomegranate seedlings with bentonite clay @ 6 kg / pit, pond silt @ 10 kg/ pit or Jalasakti @ 20 gm / pit were found most effective.
4. Application of fly ash @ 40 t/ha or 6 kg / pit increased the water holding capacity and plant nutrient availability in sandy soil of desertic region or coarse textured soils. However, before its use for agronomic purposes, it is necessary to ascertain the amount of toxic heavy metals in soils and their release pattern over time.
5. In coastal sandy soils, planting of tree species like Jhau, casurina and eucalyptus would act as a shelter belt against cyclone and reduced soil erosion. Coconut, arecanut also grow well in coastal sandy soils.

Management of Shallow, Stony and Shallow- water table soils

Shallow soils

Soil depth, in addition to soil texture and soil structure, can influence water and nutrient availability. Based on the depth of the soil profile from top to bed rock, the soils are classified as:

i) Very shallow - < 25 cm
ii) Shallow-25 to 30 cm
iii) Moderately deep-50 to 90 cm
iv) Deep-90 to 150 cm
v) Very deep-More than 150 cm

Shallow soils are mostly reddish brown to red color, poorly aggregated and developed from granite-gneiss complex. The major physical problem of the swallow soil is quick drying of the surface layer due to the presence of murrum at shallow depth restricting the capillary flow of water from the underlying layers. Seed germination and root elongation are major concern in shallow soils.Soil erosion is a serious problem in these soils.

Management

1. Planting a permanent ground cover or mulching helps to minimize erosion from surface soil.
2. Construction of ridges against the slope increases infiltration, conserve moisture in soil, increase root growth and crop yield.

3. Fruit crops like mango, jack fruit, cashew, ber, tamarind can withstand shallow soil depth.
4. Follow silvi-pastoral system of cultivation.

Stony soils

Stony soils were identified where the soil depth has been mapped as less than 45 cm to gravels. The stony soils include rounded-stony, angular-stony and fragmental soil profile materials. The stony soils are characterized by moderate to rapid permeability, low available water storage capacity and low P- retention capacity. These attributes are all indicators of stony soils with high to very high vulnerability to leach nutrients.

Stony soils contain high amounts of rock fragments and formed from lime stone, sand stone, shale, rocks or weathering resistant parent materials. The amount and type of rock fragments in surface layers can affect infiltration and water storage, which in turn influence land use and site productivity. Rock fragments influence soil hydraulic properties by affecting porosity and the tortuosity of water flow paths. With increasing the amounts of non-permeable rock fragments the area available for water flow decreases, tortuosity of the flow path increases and over all water movement can become restricted.

The areas with steep slopes and high rock fragment content, remained under forest ecosystem. The areas on hill bottom may be fit for pasture cultivation. The steep slope soils have more rock fragments, less silt, lower bulk density and higher porosity than hill soils.

Management

Stony soils are more prone to run off and leaching of water and nutrients depending on the stiffness of the slope. Management practices and land utilization pattern for forestry, pasture development, agriculture and horticulture could be taken up depending upon land situation, soil characteristics and rainfall pattern.

Stony soils in steep slopes should permanently be placed under forest species. This will helps to stabilize and improve agriculture on lands down below by minimizing soil erosion.

The soils in gentle slope can be utilized for silvi-pastoral practices including fuel-fodder-horticultural plantations. The valley bottom soils should be set for agricultural crops. Horticulture should be developed in suitable localities.

The agronomic practices and engineering measures like terracing, bunding, crop rotation, maintenance of soil fertility with cover crops and rain water management should be followed to maintain productivity.

Shallow- water table soils

Shallow- water table is a condition where the seasonal high ground water table or saturated soil is less than 3 feet from the land surface. In humid and sub humid regions, the seasonal water table is located less than 3 feet from the surface covering about 50% land mass in medium and low land situations. Shallow ground water reduces the depth of the unsaturated soil available for treatment, leading to an increased likelihood of groundwater contamination. The **Vadose zone** is further reduced when groundwater mound forms. These sites present challenges for removal of pollutants. Pollutant removal in the vadose zone is attained via biological activity, chemical degradation, adsorption of pollutants to soil and plant uptake.

Constructing a sedimentation pond achieves pollutant removal through the processes of settling of suspended solids. If the pond basin is large enough and has a long retention time, growing of vegetation will reduce the pollutant load through plant uptake. An infiltration practice removes pollutants through filtering in the unsaturated soil layer. This will help to increase the microbial activities in the layer under aerobic conditions.

References

Baver, L.D. (1959) Soil physics. John Wiley & Sons. Inc. New York. Champan& Hall Ltd. London.

USDA (1999) Soil quality test kit guide. USDA Soil Quality Institute. Washington, D.C.

Study Questions

1. What is soil crusting? Name important soil properties that encourage crust formation.
2. Briefly discuss the mechanism of crust formation.
3. What are the impacts of soil crusting? How you will overcome this problem?
4. Differentiate between compression and compaction.
5. What are the soil physical changes occurred due to puddling?
6. Briefly discuss the mechanism and problems of soil compaction in agriculture.
7. State different management practices to overcome compaction problem in crop production.
8. Soil compaction is a serious problem in rice-rice cropping system in irrigated command area – Explain.

9. Briefly discuss the crop production constraints and remedial measures of following problematic soils.
   i) Slow permeable soils
   ii) High permeable soils
   iii) Shallow soils
   iv) Shallow-water table soils
10. Explain the following?
   i) Regur soils
   ii) Arch action
   iii) Puddlability
   iv) Void ratio
   v) Compression index
   vi) Plow sole
   vii) Permeability
   viii) Vadose zone

3

Management of Calcareous and Acid Sulphate Soils

Calcareous soils

In arid regions, calcareous soils are developed from limestone or marl parent materials. The content of calcium and magnesium carbonate may be as high as 60 to 70 per cent of the total soil mass. In calcareous soil lime content may vary from a small amount in some places in the profile to appreciable amounts throughout the profile.

Calcium carbonate coatings occur in sand, silt and clay size fractions of calcareous soils. In sub soil, large amounts of lime forms hardpan under high water tables.

In general, pH of the soil increases to alkaline pH with increasing the proportion of water to soil ratio. This may happen in part from a reduction of CO_2 pressure. The dilution of salts in soil solution also permits greater hydrolysis of calcium clay. The dilution of calcium carbonate with production of hydroxyl ions may lead to higher pH as:

$$CaCO_3 + H_2O \leftrightarrows Ca^{2+} + HCO^-_3 + OH^-$$

The build- up of hydroxyl ions from the hydrolysis of calcium carbonate is prevented by adequate carbon dioxide in soils.

Reactions between calcium carbonate ions and iron may reduce iron availability to plants:

$$Fe^{2+} + CaCO_3 \leftrightarrows FeCO_3 + Ca^{2+}$$

$$4\ FeCO_3 + O_2 + Ca(HCO_3)_2 \leftrightarrows 2Fe_2(CO_3)_3 + Ca(OH)_2$$

$$Fe_2\ (CO_3)_3 + 3\ H_2O \leftrightarrows Fe_2O_3 + 3H_2CO_3$$

Since, iron is assimilated by plants as ferrous iron, moist calcareous soils would not favour iron availability.

Soils high in lime are productive and support many field crops like forage crops, corn, cotton, sugar beets, potatoes and tomatoes. Some other crops like citrus, peas, beans, suffer from acute iron deficiency. In calcareous soils, the yellowing of leaves is termed **lime-induced** chlorosis developed by iron deficiency in plant leaves. High amount of calcium and bicarbonate in soil solution are associated with chlorosis problems in calcareous soils. Yellow leaves due to lime induced chlorosis contain high amount of potassium, organic acid and soluble nitrogen. Acid soluble iron is low as compared to green leaves.

It is also observed that bicarbonate ions interfere with iron nutrition of many plants. This relationship support the hypothesis that the bicarbonate ions is an important factors in lime-induced chlorosis.

Management of calcareous soils

1. The build-up of hydroxyl ions from the hydrolysis of calcium carbonate is prevented by adequate carbon dioxide in soil.
2. In presence of excess water with low carbon dioxide pressure, calcareous soils may become excessively alkaline. The control of excess irrigation seems to be more important in management of calcareous soils. Furrow irrigation is better than basin irrigation on slaking calcareous soils. Deep irrigation may also be practiced.
3. Acidification of alkaline - calcareous soils increases plant growth and nutrient availability. However, soils containing 5% or more lime with high buffering capacity do not respond to acid treatments and increase uptake of phosphorus, iron, manganese or zinc.
4. Calcareous soils generally have low organic matter content and low nitrogen. Nitrogen fertilizer may be applied at any time from planting to crop establishment. Side dressing of nitrogen fertilizers to growing crops is an efficient method of nitrogen application. Care should be taken to avoid volatilization loss of nitrogen.
5. Phosphorus is lacking in calcareous soils. Excess application of phosphorous may lead to zinc and iron deficiency. Band placement of water soluble P at sowing is preferred to broadcasting.
6. Calcareous soils generally suffer from Fe and Zn deficiency. Zinc sulphate is an effective source of Zn. Foliar application of Zn and iron are recommended in fruit crops. Heavy application of animal manures are helpful in preventing the deficiency of Zn and Fe.

Acid Sulphate Soils

Million hectare of low-lying coastal land in tropics are either poorly productive or unsuitable for agricultures because the soils are acid sulphate or potentially acid sulphate or as the Dutch refer to those soils **Cat clay**.

Acid sulphate soils are soils that have, somewhere within a 50 cm depth, a pH below 3.5 (Entisols) or 4.0 (Inceptisols) that is directly or indirectly caused by sulphuric acid formed by oxidation of pyrite (Fe S_2). Generally acid sulphate soils are found in coastal areas where the land is inundated by salt water. These soils contain sufficient iron sulphides to produce more sulphuric acid when oxidized and cannot be neutralized by the inherent buffering capacity of soil.

Occurrence

In India, acid sulphate soils are mostly found in Kerala, Odisha, Andhra Pradesh, Tamil Nadu and West Bengal. The area covered under acid sulphate soils in Indonesia is about 2 million hectares. Occurrence of acid sulphate soils in some countries are given in table 3.2.1.

Table 3.2.1. Distribution of acid sulphate soils in selective countries.

S.No.	Country	Area (lakh ha)	Broad group
1.	Bangladesh	7.00	Sulfaquents
2.	Burma	1.80	Sulfaquents
3.	China	0.67	Sulfaquents
4.	India	3.90	Highly organic Sulfaquents
5.	Indonesia	20.00	Highly organic Sulfaquents
6.	Malaysia	1.60	Highly organic Sulfaquents
7.	Thailand	6.00	Sulfaquents
8.	Vietnam	10.00	Sulfaquents

Genesis of acid sulphate soils

Acid sulphate soils are formed by the oxidation of pyrite (FeS_2). The soil contains about 2-10 per cent of FeS_2 on volume basis. The ingradients for pyrite formation are sulphate, iron, metabolizable organic matter, sulphate reducing bacteria and aeration.

The soils and sediments most prone to becoming acid sulphate soils are formed due to inundation of land in seacoast. Sulphate is abundant in sea water. The clayey sediments of most tidal swamps contain iron oxide (FeO) which helps in pyrite formation. Mangrove vegetation provides sufficient organic matter for formation of pyrite.

Under anaerobic conditions lithotropic bacteria such as *Desulfovibrio desulfuricans* obtain oxygen for respiration through the reduction of sulphate ions producing hydrogen sulfide.

These bacteria can only survive when there is an anaerobic environment (free of Oxygen). The lack of oxygen protects both bacteria and pyrite they create.

The sulphate reducing bacteria use sulphate and produce hydrogen sulfide (H_2S) the way animals use oxygen and produce carbon dioxide. Iron in its reduce form (Fe^{2+}) reacts with hydrogen sulfide (H_2S) forms very fine grained and highly reactive crystals of iron sulfide such as pyrite.

Tropical water logged environments such as mangrove swamps or estuaries, contain higher level of pyrite than temperate climate.

The pyrite is stable until exposed to air. When the soil is drained and then aerated, the sulphide (S^{2-}) is oxidized to sulphate (SO_4^{2-}) by a combination of chemical and bacterial action, forming sulphuric acid. The magnitude of acid development depends on the amount of sulphide present in soil and the conditions and time of oxidation.

Oxidation of pyrite

After drainage, the physically unripe soil will crack and air will penetrate. In well aerated moist soil, the pyrite will oxidize to dissolve ferrous sulphate and sulphuric acid. The overall reaction in absence of bacteria is -

$$2FeS_2 + 2H_2O + 7O_2 \rightarrow 2\ FeSO_4 + 2H_2SO_4$$

The chemical oxidation of pyrite is quite slow. But the oxidation rate increases in the presence of microbe *genus Thiobacillus.*

Below pH 2.0, the bacteria *Thiobacillus ferroxidan* are the most active oxidizers and the activity build up rapidly. The biological oxidation of pyrite is given by the following reaction.

$$2S + 3O_2 + 2H_2O \xrightarrow[\text{Oxidation}]{\text{Microbial}} 2H_2SO_4$$

$$H_2S + 2O_2 \xrightarrow[\text{Oxidation}]{\text{Microbial}} H_2SO_4$$

These autotrophs derive energy from oxidation of reduced sulphur. When potentially acid soils are exposed to air in moist condition, strong acidification often occurs within a few week. Within a few months the pH falls below 3 or 4.

Buffer reaction and pH

Calcium carbonate present in aerated pyrite sediments rapidly dissolved by sulphuric acid. Sulphuric acid in excess of $CaCO_3$ reacts with clay minerals and other silicates. The reaction releases cations like aluminium, magnesium and silica from the mineral lattice. The various clay minerals tends to buffer the pH at 3-4.

Strong acidity in acid sulphate soils results toxicities of aluminium, iron, soluble salts, manganese and hydrogen sulphide (H_2S)gas. Hydrogen sulfide often formed in low land rice soils cause **akiochi** disease that prevents rice plant roots from absorbing nutrients.

Classification of acid sulphate soils

Acid sulphate soils contain sulphuric horizon which has a pH of the 1:1 soil: water ratio of less than 3.5. Sulphihemists, Sulphohemists, Sulphaquents and Sulphaquepts great groups are included in acid sulphate soils.

The acid sulphate soils are classified into

i) **Potential acid sulphatesoils** which contain sulfidic materials. Sulfidic materials are water logged materials or organic minerals with > 0.75 % sulphur. Potential acid sulphate soils may be-

a) **Sulfaquents**-Aquents with sulfidic materials within 50 cm of the mineral soil surface.

b) **SulficFluvaquent**-Fluvaquents with sulfidic materials between 50-110 cm soil depth.

ii) **Actual acid sulphate soils** contain sulphuric horizon. Sulphuric horizon is composed of mineral or organic soil material with pH < 3.5 and yellow jarosite mottles.

Identification of acid sulphate soils

Identification of acid sulphate soils in the field is easy in most cases because of the presence of the typical straw-yellow jarosite$KFe_3(SO_4)_2(OH)_6$ mottles at a depth of 50 cm. However, pH can be measured for proper identification.

Chemical changes in acid sulphate soils upon flooding

Strong acidity in acid sulphate soils results toxicities of aluminium and iron, soluble salts, manganese and hydrogen sulphide (H_2S) gas. Hydrogen sulphide often formed in low land rice soils causes **akiochi** disease that prevents rice plant roots from absorbing nutrients.

The chemical changes occurred after flooding of acid sulphate soils are

i) The pH of most flooded acid sulphate soils increases slowly (compared to normal soils) and rarely exceeds 6.0 after six months of submergence. The increase in pH upon flooding is mainly due to reduction of ferric oxides to dissolved Fe^{2+}, a process that consumes acidity.

ii) There was slow decrease in redox potential.

iii) Dissolved Fe^{2+} increases to several hundred to several thousand ppm. When pH is increased to neutrality after liming, Fe^{2+} shows little tendency to decrease.

iv) Dissolved aluminium decreases as the pH rises.

v) Dissolved SO^{2-}_4 increases as the pH increases due to soil reduction.

Adverse effect on ecology and soil properties

Impact on ecology

Formation of acid sulphate soils will leads to i) Ecological damage to aquatic species ii) Effects on estuarine fisheries iii) Contamination of ground water and surface water with arsenic, aluminium and other metals. iv) Reduction in agricultural productivity through metal contamination.

Effects on soil properties and rice crop

A good rice crop can be grown in potentially acid mangrove marshes where sufficient fresh water is available otherwise, the rice yield is low due to salinization and acidfication during droughts.

Other common soil related problems in acid sulphate soils are low pH, toxicities of aluminium, iron, hydrogen sulfide, carbon dioxide, salt injury, toxicity of organic acid, deficiency of phosphorus and micronutrients and low nutrient status.

i) High aluminium level affects cell division, disrupts certain enzyme systems and hampers uptake of P, Ca and K.

ii) Dissolved iron in excess of 300-400 ppm is toxic to rice. This problem occurs for many months in flooded sulfaquents with high organic matter.

iii) Acid sulphate soils in tidal areas are often affected by soil salinity.

iv) Even in absence of Fe and Al-toxicity and salinity, P- deficiency is the most disorder of the rice plant on acid sulphate soils.

Management

Management techniques are extremely variable and depend on many specific factors viz. the extent of acid formation, the thickness of the sulphide layer, possibilities of leaching or draining the land etc.The dominant adverse soil factors are iron toxicity and sometimes P-deficiency in Sulfaquepts and sulficTropaquepts. Aluminium toxicity may be a problem especially in deep water areas.

1. Agro Technology

Application of lime (in sulfaquepts), P fertilizers (inSulficTropaquepts) and the use of suitable rice varieties were found beneficial in increasing soil productivity.

2. Water Management

Good water management especially in tidal areas with sulfaquents and in swamps with highly acid sulfaquepts are important in improving rice productivity.

3. Controlling Water Table

By keeping the water table above the pyrite subsoil and by preventing prolonged drying of the surface soil, the pH, which controls all other adverse factors, is kept in a safe range.

4. Leaching

If these soils are leached during early stage of acidification, lime requirement values decreased. Leaching, however, is difficult because of high water table commonly found in this type of soils and low permeability of clay. Sea water is sometimes available for leaching.

5. System of rice cultivation

For success of rice crop, following steps are recommended:

- Before transplanting, fill the field with water and drain it.
- Repeat this process several times in order to leach acid and toxic materials from the soil.
- Add lime (slaked lime, marl, lime stone) to decrease soil acidity.
- After adding lime, leave 7 to 10 days. Then add phosphatic fertilizer before planting of rice seedlings.
- Add silica for higher yield.
- Transplant aged (30-35 days) seedlings with 3-5 seedlings / hill.

Study Questions

1. Lime-induced chlorosis is a major problem in citrus – Why?
2. Under what conditions the calcareous soils become excessively alkaline?
3. Why the mangrove environment contain higher amount of pyrite than temperate climate?
4. Name different ingradients required for formation of pyrite.

5. Where are acid sulphate soils found?
6. Briefly discuss the genesis of acid sulphate soils?
7. How the pH of acid sulphate soil is buffered at 3-4.
8. What are the chemical changes occurred in acid sulphate soils on flooding?
9. Suggest some management practices for rice in acid sulphate soils.
10. Is it possible to prevent the oxidation of pyrite in mangrove forest ecosystem? Suggest some possible methods.
11. Does acidification of alkaline–calcareous soils is helpful in nutrient availability?

4

Iron and Aluminium Toxicity

Iron Toxic Soils

Lowland rice is being cultivated on approximately 128 million hectares of either irrigated or rainfed land out of which 100 million hectares suffer due to nutrient deficiency or toxicities. Among the toxicities, iron toxicity is well recognized as a nutritional disorder in low land rice production. Iron toxicity is a syndrome of disorder associated with high concentration of Fe^{2+} in soil solution. A wide range of soil types show iron toxicity problem including acid sulphate soils, acid clay soils, peat soil and valley bottom soil. During monsoon rains in the wet season, soluble iron (Fe^{2+}) migrate to the surface and its iron content increases to subtoxic or toxic levels.

In rolling topography and with heavy monsoon downpour there is considerable lateral flow of iron containing water from the adjacent uplands to mid – and low lands. This increases the iron content to toxic levels. Poorly drained sandy soils in valleys receiving interflow water from adjacent uplands with laterite horizon (Plinthudults) showed iron toxicity in Srilanka and Kerala and Odisha in India. These soil groups come under the taxonomic orders Alfisols, Ultisols and Oxisols occurring in association with Entisols and Inceptisols. The iron concentration (Fe^{2+}) in soil solution that affect low land rice yield ranges from 10 to > 2000 mg L^{-1} Iron toxicity induced yield reduction which is associated with poor nutrients status (deficiency of P, K and Zn) and H_2S toxicities.

Bronzing systems of rice observed when the iron concentration in rooting medium ranges from 300 to 500 mg L^{-1}. With low nutrient status or with respiration inhibitors (H_2S), iron concentration of 30 mg L^{-1} may be toxic to rice plants.

Soil Reduction and Iron Toxicity

Soil reduction influences iron toxicity in at least three ways:

i) Change in concentration of dissolve Fe^{2+}

ii) Formation of reducing organic substances

iii) Production of dissolved sulfide

The general characteristic of soil showing iron toxicity are presence of high amounts of reducible iron, low pH, low CEC and exchangeable potassium content. These may be associated with P and Zn deficiency and H_2S toxicity. Iron toxicity occurs under water logged situation with poor drainage. Since oxygen diffuses in air about 10^3 - 10^4 times faster than in water or water saturated soils, oxygen is depleted rapidly by respiration of micro-organisms and plant roots in water logged condition. With depletion of oxygen, NO^-_3, Mn^{4+}, Fe^{3+} and SO^{2-}_4 act as electron acceptors for microbial respiration and become reduced in a flooded rice soils. Within a few days after flooding the reduction of Fe^{3+} begins at a redox potential of 170 to -180 mv at 25^{0}C. The reduction reaction is given by:

$$Fe(OH)_3 + 3H^+ + e^- \leftrightarrows Fe^{2+} + 3H_2O$$

It is often found that the concentration of ferrous iron in the soil solution rises rapidly to a maximum and then slowly falls as the reaction proceeds due to precipitation of $Fe_3(OH)_8$ at hgher pH.

The amount of Fe^{2+} increases with quantity of decomposable organic matter, the temperature and the absence of compounds with a higher oxidation state than Fe(III)-oxide. It also increases with the duration of submergence and attained the peak values at 2-8 weeks after flooding and remained constant thereafter. Under waterlogged rice soils, three distinct layers have been observed i) A thin oxic surface layer (2 to 10 mm) , ii) The reduced puddle bulk soil and iii) The oxic rhizosphere soil.

Highest Fe^{2+} concentrations are found at 2-15 cm soil depth. The concentration of Fe^{2+} is less in clay than in sandy soils because of retention of iron on clay – mineral surfaces. The iron toxicity problem occurs frequently in lowland soils with kaolinite than with smectite. The concentration of soluble iron in acid soils varies from 100 to 1000 mg L^{-1} where as in acid sulphate soils it may rise to more than 5000 mgL^{-1}. The critical concentration of Fe for low land rice is considered as 300 mg L^{-1} but, varies between 20 to 2500 mg L^{-1} depending on the rice cultivars and site characteristics. In reduced soils, sulfate is unstable and is reduced to dissolved sulfide. Dissolved sulfide is harmful to rice and aggravate iron toxicity.

Iron Toxicity and Nutrient Availability

Formation of ferrous and manganous ions under water logged condition can take part in cation exchange and will increase the amounts of exchangeable Ca, Mg and K that come into solution. In the process of reduction of Fe (III) oxides,P and to a lesser extent Mo can be released, but the availability of toxic Al^{3+} may also increases. Consequently, large quintiles of Al^{3+} can also be taken up along with iron by the rice plant. In absence of fertilizer-P, the P released from iron phosphate due to reduction of Fe^{3+} is the dominant source of P for rice. With increasing the availability of P, the Fe^{2+} formed by the reduction of Fe oxides shows antagonistic interactions with other cations such as Mn, Zn and K causing deficiencies of these cations in rice. Iron oxides have a strong Zn binding capacity. When soils are flooded, Zn becomes available in the process of iron oxide reduction. However, reduced iron can also exert a direct antagonistic effect on Zn uptake. The uptake of K is also affected by excess of Fe^{2+} in soil solution. Several experimental findings support the ameliorative effects of fertilizer P, K, Zn and Mn application on performance of lowland rice under iron toxic conditions.

Relationship of DTPA-Fe with other Soil Properties

The research results of the experiments conducted in the Department of soil Science and Agricultural Chemistry, Orissa University of Agriculture & Technology, Bhubaneswar showed that DTPA-Fe had negative and significant correlations with pH, K, Ca, Mg and Mn and Clay content but had a positive correlation with Zn. The multiple regression equation is represented by –

DTPA-Fe = 99.8 + 128.3 silt + 257.0 clay + 459.4 pH + 199.7 OC – 1.1 N + 15.9 P – 2.0 K – 119.8 Ca – 69.5 Mg + 196.7 Zn – 1.2 Mn+13.1 Cu (R^2 = 0.850)

All these factors collectively contribute 85% towards iron toxicity. The major contributors were soil pH (45.8%), clay (14.3%), organic carbon (8.6%) and silt (3.5%). Clay and organic carbon have relatively lower contribution since the soils are predominantly sandy or sandy loam with low organic matter content.

Iron toxicity symptoms

Even though iron is an essential plant nutrient and strongly promotes rice root development at low concentrations, excess ferrous iron in soil retards rice growth. Iron toxicity symptoms manifest in soil, leaf, root and grain. Brick red color oily scum floating on surface water adjacent to bunds is a common feature of iron toxic soils. The bronzing symptoms starts with purplish or reddish brown mottling or some times, a yellowish or orange discoloration spreads downwards from the tips of older leaves followed by drying of leaves. Roots are scanty, short, coarse and often dark brown and white roots are few or absent. Toxicity during vegetative stages is associated with stunted growth, extremely limited tillering and short height. Both the tiller formation and the number of effective tillers reduced. Toxicity at early reproductive stage is associated with fewer panicles per hills, an increase in spikelet sterility and delayed flowering to maturity by 20-25 days. Under severe cases, the seed coats became tinged with brown spots.

The yield of rice significantly correlated with iron toxicity symptoms. The average yield loss due to iron toxicity ranged from 12 to 35%. However, toxicity at tillering stage can strongly affect plant growth resulting complete yield loss.

Criteria for evaluation of iron toxicity

Three criteria have been used to define iron toxic levels as:

i) Plant growth

ii) Degree of bronzing

iii) Iron content of plant

Rice grows well in 0.1–1.0 mg L^{-1} ferrous iron. Bronzing symptoms appear generally at higher Fe^{2+} concentration (>30 mg L^{-1}).The yield of rice is affected with degree of bronzing. The iron content in plant of a given variety at a given time and place usually correlate well with degree of bronzing. The iron content of a bronzing plant vary from 110 to 1100 mg kg^{-1} but, healthy plants sometimes also contain more than 1000-1500 ppm iron. Therefore it is impossible to define a applicable critical iron content in plant tissue. To confirm iron toxicity, a comparison of the iron contents in the leaf blades of affected and healthy plants from the same field may be more useful.

Bronzing often develops at the maximum tillering and panicle initiation stages, but may be observed at any growth stages. Even if it occurs at flowering stage.

Silica has a positive effect on the oxidizing capacity of the rice roots and depress the degree of bronzing by reducing the uptake of iron at high Fe^{2+} concentration. Low concentration of K, Ca, Mg, P and Mn in solution increases the uptake of iron at high Fe^{2+} concentration. Low K and P apparently enhance iron-toxicity through decreased oxidizing capacity of the roots. In contrast, N deficiency does not result in higher iron uptake. Low pH and high salinity due to NaCl or $MgCl_2$ also increase iron uptake and degree of bronzing in plant.

Management of iron toxicity

A number of ameliorate practices are recommended for iron toxic soils. These are discussed in following paragraph.

1. Iron exclusion mechanism

Reduction of Fe^{3+} to Fe^{2+} ions either in situ in the soil or interflow can be reduced by increasing the root oxidizing power. The presence of **aerenchyma** (gas conducting issues) in the rice roots provide a pathway for molecular oxygen transport from shoots to roots. Because of this mechanism the iron tolerant rice varieties could increase their oxidizing power and improve their tolerance to iron toxicity. The aerenchyma may occupy 20-50% of the total root volume of flooded rice. The extent of aerenchyma formation well correlated with the rate of ethylene formation in root tissue. Young secondary roots and root tips diffuse oxygen at the highest rates into the rhizosphere. Ethylene-induced aerenchyma formation starts in 2-4 week in old plants and the oxidation power of the root is highest at maximum tillering stage.

2. Avoiding a rapid drop in Eh

Under flooded condition the **Redox Potential (Eh)** of soil falls rapidly and reduction of Fe^{3+} begins at a redox potential of 170 to -180 mv.

High level of reducible iron combined with large quantities of organic matter increase the severity of iron toxicity symptoms. Addition of easily decomposable organic matter (green manure) should be avoided, particularly in alluvial clays and acid sulphate soils to reduce iron toxicity.

3. Application of essential plant nutrients

Application of essential plant nutrients counteract the adverse effect of iron toxicity by competing with Fe^{2+} in crop uptake or ion-adsorption sites at the root surface or influencing tolerance mechanisms. Application of potassium regulate iron uptake in rice plant.

Incorporation of rice straw can serve as an alternate source of potassium and silica. Zinc application increases the tolerance of rice to iron toxicity.

Depending on the soil type, application of P, K, Ca, Mg and Zn alone or in combinations reduce bronzing symptoms and increase rice yield. Surface application of 100-200 kg MnO_2 per hectare at planting reduces iron toxicity. Application of silica increases oxidizing power of rice roots.

4. Agronomic practices

Cultivation of long duration local rice varieties or salt tolerant rice varieties, delay in transplanting by 3-4 weeks after flooding to overcome the peak concentration of Fe^{2+} ions, dense planting of rice and drainage if possible are some of the agronomic practices recommended to overcome iron toxicity problem.

Aluminium Toxicity

Aluminium (Al) toxicity is the most common cause of growth reduction on strongly acid soils. About 40% of the World's arable soils are acidic and subjected to aluminium toxicity at pH < 5or even 5.5in kaolinite soils. Aluminium is the most abundant metal in the earth's crust, but its solubility is low in neutral to alkaline pH. However, below about pH 5.0, aluminium solubility increases rapidly and may reach toxic concentrations. The appearance of aluminium toxicity symptoms in plants is not well correlated with the Al content in either the solid phase or the soil solution. Exchangeable Al, Al- saturation and the activity of total Al in soil solution are not suitable indicators of Al toxicity, whereas the soil pH and the Al species present in soil solution are thought to be more important. Some studies indicate that only monomeric Al species are toxic. Soluble organo-Al complexes and Al-F and Al – SO_4 complexes are less phytotoxic.

Toxicity Symptoms

Reduction in root development is the primary effect of Al toxicity. Roots appear short and thickened with short laterals and may be discolored yellow to brown. Root hair development is suppressed. Where the Al concentration increases with depth, the root penetration may be restricted, resulting in a very shallow root system. Poor plant growth with stunted appearance and water stress symptoms are common in Al toxic soils. As Al in the soil solution inhibits the uptake of Ca and Mg by roots, the symptoms of Ca and Mg deficiency may develop. The solubility of P is reduced in the presence of high concentration of Al, so that P deficiency is often associated with Al toxicity.

A decrease in Al toxicity with increasing Ca concentration in solution has been reported. The concentration of soluble Ca at which Al toxicity becomes evident depend on plant species and amount of Al present. The Al concentration in soil solution at which symptoms of Al toxicity appear in crop species varies from 50-3000 micro mole per litre. The Al toxicity in plants occur at low Ca/Al molar ratio varying from 0.2 to 2.5.

The phytotoxic form of Al

Aluminium hydrolyses in soil solution such that trivalent Al species Al^{3+}, dominates in acidic condition (pH< 5.0), whereas the $Al(OH)^{2+}$ and $Al(OH)^{+}_{2}$ species are formed as the pH increases. At neutral pH the solid Al $(OH)_3$ or gibbsite occurs whereas, Al $(0H)^{-}_{4}$ or aluminate dominates in alkaline conditions. Most of these monomeric Al cations bind to organic acids and inorganic ligands such as $PO^{3}{}_{4}^{-}$, $SO^{2-}{}_{4}$, F^{-}, proteins and lipids. Among all these species, Al^{3+} is the major phytotoxic form of Al. The root apex (root cap, meristem and elongation zone) accumulates more Al and cause greater physical damage than the mature root tissues.

Soil and plant tests

Aluminium toxicity is to be judged by testing the soil. Soil with pH below 5.0 (Soil: water ratio 1:5) show Al toxicity and soils with pH less than 4.5 severely reduce crop production.

Tests for exchangeable Al and Al saturation capacity may be useful. The tolerance of crops to solution Al is greatly increased in presence of high concentration of Ca and it is necessary to measure Ca availability to predict crop response to Al.

Management

Aluminium toxicity can be corrected with rise in soil pH, through incorporation of lime or dolomite in soil. Maintenance of high soil organic matter through incorporation of crop residues, cover cropping or mulching will bind free Al into non-toxic complexes. Phosphorus forms insoluble complexes with Al. Addition of P fertilizers decrease solution Al concentrations.

Study Questions

1. Where do you notice iron toxic soil?
2. How do the soil reduction process influence iron toxicity?
3. Why do the non-paddy crops grown in iron rich soils not suffer due to iron toxicity?
4. What is the relationship between ferrous iron concentration and days of submergence in iron toxic soils?
5. In addition to iron toxic problem, rice also suffers due to nutrient deficiency. Does the iron toxic soils are deficient in nutrients- Explain?
6. Briefly discuss the iron toxicity systems in rice crop.
7. Mention some agronomic practices to overcome iron toxicity problems.
8. Name different aluminium species that occur in acidic, neutral and alkaline pH conditions. Which species is toxic?
9. State the toxic effects of aluminium on plant roots.
10. Explain the lime and aluminium interactions in low pH soil.

5

Problems and Management of Eroded Soils

Soil erosion refers to detachment and transport of soil and soil particles from one place to other by the action of water, wind or gravity. Soil erosion consists of (i) detachment of soil particles from surface (ii) their transport from one place to other and (iii) deposition in another place.

Soil erosion are of two types: **Natural erosion and accelerated erosion**. The natural or **geological erosion** is steady, nondestructive and slow process of nature. On the contrary the accelerated erosion caused due to human activities like cutting of forest, intensive cropping, construction of roads and building is very rapid and destructive. Surface one inch of soil which is formed in thousand years can be lost within one year due to accelerated erosion.

In India, out of total 175 million hectares of degraded land, soil erosion due to water and wind accounts for 150 million hectare. Soil erosion due to shifting cultivation, salinity, alkalinity, water logging and shifting of water courses account for 25 million hectare.

Average soil loss from various land surfaces in India has been estimated to be 5334 million tonnes per year out of which 29 percent is permanently lost to sea, 10 percent is deposited in reservoirs and 61 percent is transported from one place to other. The plant nutrients lost through 5334 million tonnes of eroded soils was estimated to be 8.4 million tonnes which includes 2.5 million tonnes of N, 3.3 million tonnes of P and 2.6 million tonnes of K.

Siltation of reservoirs through water erosion is a major concern in India. This accelerated silting has considerably reduced the estimated life span of major reservoirs. The rate of siltation depends on the catchment area, soil erodability, topography and rainfall pattern. Larger the catchment area, smaller is the siltation load in the reservoirs because, the coarse fraction of eroded particles may be settle down in the stream valley before the water enters into the reservoir.

Causes of water erosion

i) Rain drops impact

Hitting action of rain drops on soil surface has three impacts - (i) It detaches soil particles (ii) Destroys granulation and (iii) Splashes soil particles. About 95% of soil is splashed by hitting action of rain drops and run off water erodes less than 5% of the soil. On an acre of land the total energy of rain drops is calculated to be roughly 100 HP during rainfall of 0.25 cm h^{-1} and 250 HP at 5.0 cm h^{-1}. On steep slopes, the total energy of rain drops may increase. The direct impact of falling rain drops destroys granulation, decreases infiltration resulting run off.

Runoff may occurs without erosion but erosion never occurs without run off. Runoff that causes erosion depends on amount, duration, intensity and frequency of rainfall. A rainfall of 5cm day^{-1} always cause run off.

ii) Topography

The velocity of run off increases with increasing the steepness of slope. It was reported that a four-time increase in the degree of slope, doubles the velocity of flowing water which increase its erosive power by four times and the silt carrying capacity by 32 times. The soil erosion is also influenced by slope characteristics like steepness of slope, slope length and micro topography. The length of slope is the distance from the beginning of runoff to the point of deposition. Data given in the table 5.1 shows the intensity of run off and soil loss at different slope percents.

iii) Soil Properties

Soil properties such as texture, structure, organic matter content, nature of clay and amount of salt present influence soil erodibility. Fine texture soils and alkali soil are more prone to soil erosions as compared to soils with high silica / sesquioxide ratio. Coarse textured sandy soils are more prone to wind erosion. Soils with higher absorptive and water retention capacity are les erodible because of low runoff rate.

Table 5.1. Effect of slope on soil, water and nutrient loss.

Slope (%)	Soil loss (tonnes ha^{-1})	Run off loss (%)	Average nutrient loss (Kg ha^{-1})		
			N	P_2O_5	K_2O
0.5	3.0	34	0.7	2.3	11.7
1.5	9.4	42	1.4	3.8	17.1
3.0	13.6	51	4.6	9.0	43.1

Source: Tejwani and Bhardwaj (1982)

Types of Soil Erosion

i) Splash Erosion

Splash erosion is considered as first stage of soil erosion. When rain drop hits the ground surface, soil particles became loose and splashed due to its force.

The forces and soil properties that influences splashing are raindrop velocity and soil characteristics such as particle size, infiltration, moisture content and land features like roughness, surface slope. The erosive capacity of falling rain drop depends on its diameter and distance of fall. Larger size and higher velocity of rain drop have greater kinetic energy and cause more erosion. Raindrop action can be cumulative, initially breaks the soil aggregation and later throw them out. On flat surface, the splashed particles tends to scatter in all directions but in rolling topography, a major portion of the splash moves down the hill. Detachment of soil particles is difficult in case of finer clay because of its strong binding effect. Presence of organic matter further helps in binding of the particles. A saturated soil weaken the cohesive force, reduce seepage and encourage stronger splashing.

The amount of soil loss due to splash erosion is given by:

S = kd vi where,

S -amount of soil splashed in 30 minutes, k - soil constant, d- drop diameter (mm), v - drop velocity ft/sec and i - rainfall intensity, inches / hr.

ii) Sheet Erosion

Movement of soil by rain drop splash is the primary cause of sheet erosion. Sheet erosion refers to uniform removal of soil from all parts of the area having a similar degree of slope. Sheet erosion occurs as a result of kinetic energy associated raindrop impact and runoff. Red soils and black soils are very prone to sheet erosion. Annual soil loss due to sheet erosion may vary from 11 to 43 tonnes ha^{-1} along with 10 to 30 per cent run off loss.

iii) Rill Erosion

Rill erosion is a serious problem in laterite soils because of high precipitation. In rill erosion, the soils washed out by runoff water along the slope through small – finger like channels which can be smoothed out completely by normal cultivation. During heavy rainfall, run off water is quickly concentrated in micro rills which in turn flows through a larger rill and finally discharge to an existing channel. This type of soil erosion is regarded as a transition stage between sheet and gully erosion.

iv) Gully erosion

Gully erosion is most spectacular in 4 million ha land in India. Gully erosion is an advance stage of sheet erosion which so deep that the ground cannot be smoothed out by ordinary tillage operations. Rills more than 30 cm depth are called gullies.

An advance stage of gully results into ravine which are more than 16 to 33 m deep. Gully erosion depends on several factors viz. run off producing characteristics of the watershed, drainage area, soil characteristics, slope and vegetative cover. Four different stages of gully development are recognized as:

i) **Formation stage** (advance stage of rill erosion)

ii) **Development stage** (exposure of parent material),

iii) **Healing stage** (vegetation starts growing in the gully) and

iv) **Stabilization stage** (sufficient vegetation cover develops over the gully surface).

Harmful effects of water erosion

Water erosion causes various damages to the lands as follows

i) Loss of top soil

ii) Accumulation of coarse soil materials on other productive lands

iii) Loss of plant nutrients

iv) Siltation of water reservoirs

v) Frequent floods and loss of crops

Estimation of soil loss due to water erosion

The soil loss due to water erosion is estimated by using universal soil loss equation as:

Xa = RKLSCP where,

Xa - Average annual soil loss t/ acre

R - Rainfall factor

K - Soil erodibility factor

L - Slope length factor

S - Slope steepness factor

C - Cropping – management factor

P - Conservation practices factor

Rainfall factor (R)

The rainfall factor R, is the product of kinetic energy (E) of the storm and maximum rainfall intensity over 30 minute (I) called as rainfall **Erosion Index (EI).** The rainfall factor R is expressed as:

R = Total rainfall kinetic energy (E) x I_{30} / 100 where,

I_{30} - 30 minute rainfall intensity for each storm and

E = 916 + 331 log i, i> 0 where,

E - Kinetic energy (ft-t/acre inch of rain)

i - rainfall intensity, inches / hour

The EI is calculated for an individual storm. The values of individual storms can be summed to obtain seasonal or annual soil loss. The EI must be modified to evaluate R under two conditions – (i) Run off due to snow melt and (2) Run off due to hurricane storm in coastal plain.

Soil eordibility factor, K

Soil erodibility factor K, is influenced by soil physical and chemical properties affecting detachability, transportability and infiltration rate of soil. These factors are studied separately in field and laboratory conditions.

Based on the available data, **Olson (1961)** suggested to determine K for a standard plot (22 m long with 9% slope) as:

$$K = \frac{Xa^1}{\text{Total EI}} \text{ Where,}$$

Xa^1 - expected soil loss from a standard plot.

Xa^1 Can be calculated as:

$$X^1 = \frac{Xo}{LSCP} \text{ Where,}$$

X_0 = observed soil loss from a standard plot

Topographic factor LS

The term LS is collectively called as topographic factor. The LS factor can be calculated as:

$L = (l/73)^{0.5}$ and

$$S = \frac{0.043S^2 + 0.30s + 0.43}{6.613}$$

Where, L - slope length (ft) and s - steepness of slope (%).

Slope length is the distance from the point of origin of run off to the point of deposition where run off enters into a well -defined channel.

Cropping – management factor C

The cropping management factor C is the ratio of soil loss for a given condition to soil loss from a continuous fallow on identical soil, slope and under the same rainfall. The factor C is influenced by soil cover, cropping pattern, length of growing season and tillage operations. The value of C varies from 0.01 to 1.

The conservation practice factor, P

The conservation practice factor P is the ratio of soil loss for a given practice to that for up and down the slope farming. The conservation practice factor for various practices such as contouring, strip cropping and terracing are given in the table 5.2.

Table 5.2. Recommended conservation practice factor P.

Percent slope	P values		Terracing and contouring
	Contouring	**Contour* strip cropping**	
1.1-2.0.	0.6	0.30	-
2.1-7.0	0.5	0.25	0.10
7.1-12.0	0.6	0.30	0.12
12.1-18.0	0.8	0.40	0.16
18.1-24.0	0.9	0.45	-

* A system using 4-year rotation of corn, small grain, meadow.
Source: Smith and Wischmeir (1962)

Water erosion control measures

Basic principles

Soil erosion by water occurs at two stages

i) Detachment caused due to hitting action of rain drop

ii) Transport of eroded particles through run off.

Therefore soil erosion by water can be reduced by preventing detachment and transport of the particles.

The basic principles in minimizing water erosion are:

i) Soil protection from rainfall

The hitting action of rain drop can be reduced by raising a cover crop or covering the surface with straw or leaf mulch. The vegetative cover crop absorb the kinetic energy of rain drops and reduces its beating action so that breakdown of granulation decreases. A close growing crop slows down the velocity of flowing water.

ii) Increasing infiltration capacity of soil

Run off occurs when rainfall amounts exceeds infiltration of water in soils. The factors that reduces infiltration rate increases the run off. The infiltration rate is mostly affected by type of crop, soil texture, soil structure and rain fall intensity, frequency of rainfall, surface conditions and tillage practices. Grasses, deep rooted crops generally improves soil structure and increases infiltration capacity. Soil compaction reduces infiltration rate.

iii) Control of surface run off

Eroded soil particles are transported by flowing water in downstream. The damage caused by surface runoff can be reduced by reducing concentration of runoff and its velocity by controlling the length of slope and constructing mechanical barriers on soil surface.

iv) Safe disposal of surface runoff

The excess run off from the land surface can be stored in farm ponds or water harvesting structures.

Control measures

Mechanical control measures

i) Contour bunds

The runoff water can be controlled or minimized by constructing bunds across the slope of the land. Contour bund is the most common measure for soil and water conservation in regions having less than 75 cm of annual rainfall, soil depth more than 20 cm and land slope less than 7% with good infiltration capacity.

Contour bunds are constructed across the slope of the land on contour lines so that the long slope is cut into a series of small ones and each contour bunds will act as a barrier to water flow. Contour bunds will decrease the length of slope and reduce volume of flow.

ii) Terracing

Terracing refers to construction of broad channels across the slope of the rolling land. Terracing reduces the length of slope, reduce soil erosion and retain run off water.

Terraces are classified under two types such as (a) broad base terraces and (b) bench terraces.

a) Broad base terrace

A broad base terrace is a broad surface channel or embankment constructed across the slope of sloppy lands having less than 10% slope. In high rain fall areas, the lands having 3-10% slope with low permeability, channel type broad base terraces are constructed to remove excess water. In low rainfall areas with less than 6% slope, level type or ridge type broad base terraces are constructed to retain entire rainwater without flowing to downstream.

b) Bench terrace

Bench terraces are flat beds, constructed on hills (20-30% slope) across the slope. The height of terrace should not be more than one meter and width of the terrace depends on the degree of slope. The bench terrace may be level, slopping inward or slopping outward bench terrace. Level terraces are recommended in average rainfall areas. In slopping in ward terrace, excess water is drainout as quickly as possible through an inward channel. In low rainfall areas with shallow soil depth, slopping outward terraces are recommended to reduce the existing steep slope to mild slope.

iii) Half -moon terrace

Construction of semi-circular beds of appropriate diameter in steep slopes are recommended for fruit trees or other plantation crops.

iv) Contour trenches

Contour trenches are dug out in bottom hill lands for plantation of trees and fruit plants. The dugout earth from a trench are put on the lower side of trench which acts as a barrier to store sediments and run-off water. These trenches improve water storage capacity and favour quick growth.

v) Grassed water ways

Growing of grasses on the run off route or natural drainage line bind the soil, reduces erosive velocity of run off and reduces soil and nutrient loss. Grass water ways allow excess water to flow through stream by filtering the sediments. However, accumulation of excess sediments may reduce the effectiveness of grass waterways.

vi) Water harvesting structures

Water harvesting structures are constructed down the slope to store excess water in a watershed. The harvested water can be used for irrigation and pisciculture.

Biological control measures

Erosion control measures through vegetation and adoption of agronomic practices is called biological control measures. Experimental findings revealed that a grass cover crop is five times more effective than a bare soil. The following biological measures are in vague in erosion control.

i) Contour farming

Contour farming refers to contour cultivation or apply all treatments on contours in crop production. Contour cultivation can reduce the velocity of overflow. In low rainfall areas, each contour line acts as a barrier to the flow of water. In high rainfall areas, a part of rain water is retained allowing only small volume for surface runoff.

Contouring reduces soil loss irrespective of length and degree of slope. On 4 - 6% slope with proper crop rotation, it controls erosion up to 100 m length. In steeper slopes, contour is effective upto 60 m on 8% slope, 30 m on 10% slope, 25 m on 12% slope and 30 m for steeper slope.

ii) Strip cropping

In strip cropping, crops are grown in strips or bands at right angle to the slope of the land. The strip cropping slow down the velocity of runoff water and increase infiltration rate.

There are 3 types of strip cropping such as

i) Contour strip cropping,

ii) Field strip cropping and

iii) Buffer strip cropping.

In contour strip cropping, alternate rows of erosion permitting and resisting crops are grown across the slope in regular long and narrow strips of variable width.

In field strip cropping, crops are planted in parallel strips on soils of high permeability with uniform slopes.

In buffer strip cropping, a strip of permanent cover crops (grasses) is recommended to take care of critical slopes in fields under contour strip cropping.

iii) Crop rotation

The crops in a rotation includes cereals, legumes, grasses or legume-grasses mixture. Legumes in a crop rotation improves infiltration, add organic matter and nitrogen and improves soil structure. The dense and fibrous root systems of grasses bind soil particles and create good soil structure. The legumes or grasses cover ground quickly and protect the surface from hitting action of rain drop.

Soil protecting crops such as wheat, soya, oat, barley in a cropping system allow some degree of erosion but less depletion as compared to soil depleting crops such as jowar, bajra, maize, cotton and vegetables.

The soil depleting crops are mostly erect, do not cover ground fully, possess shallow and fibrous root systems, do not hold soil together and require frequent tillage operations which make the soil more prone to erosion.

iv) Maintenance of soil fertility

A high fertility soil allows quick crop establishment than on depleted soil. A vigorous crop reduces hitting action of rain drops, reduces splash erosion, retard velocity of runoff water and bind soil particles. Application of organic manure and recommended dose of fertilizer, growing of green manure crops maintain soil fertility. Addition of organic manures improve soil structure and increase infiltration and percolation.

v) Agro Forestry

Growing of multipurpose tree species along with crops in eroded soils reduces soil loss and satisfy farmer's need (Table 5.3).

Table 5.3. Run off and soil loss under different agro forestry system.

Land use	Run off (%)	Soil loss (tonnes / ha)
Maize	18.3	17.70
Maize + Subabul	8.9	5.00
Maize + Eucalyptus	3.6	0.91
Grass + Subabul	0.6	0.13
Subabul	0.4	0.04
Grass + Eucalyptus	0.1	0.02
Eucalyptus	0.1	0.01

Source: Narain *et al.* (1994)

Wind Erosion

In India, 28.6 million ha of area (about 8.7% of the total geographical area of the country) in arid regions are affected due to wind erosion. Low rainfall (annual rainfall 100-500 mm), low humidity, strong wind, high solar radiation and single grained sand particles accelerate wind erosion. Wind takes up soil from one place and deposits another. The soil particles first **loosened, rolled or bounced** by wind along the ground surface. The ability of wind to move and pick up soil particles varies square to cube of the velocity. The dislodged soil particles move by suspension, saltation and surface creep usually occurring simultaneously.

Transportation of very fine dust particles (0.05 – 0.1 mm dia) by turbulent wind takes place due to suspension. The particles move parallel to the ground and then upward. The wind velocity more than 3 km /h is capable of lifting silt and very fine sand particles to a height of 3-4.5 km. Suspension may transport 3 to 36% of soil loss by wind erosion.

The fine grained soil particles ranging from 0.1 to 0.5 mm dia are moved by saltation. Saltation is caused by the direct pressure of wind on soil particles and their collision with other particles. The particles bounce, rise vertically, traversed and falls due to force of gravity. Wind erosion studies showed that about 50-75 % of particles are moved by saltation.

Surface creep refers to the movement of soil particles touching the ground. Breakdown of non-erodible soil particles occurs due to impact of moving particles. Movement of particles through surface creep is the outcome of the kinetic energy derived due to impact of saltation. Quartz grain particles of 0.5 to 1.0 mm dia are transported by surface creep. Laboratory studies have shown that about 7 to 15 % of soil may be transported by surface creep.

Chepil (1945) reported the relative proportion of soil transport by suspension, saltation and surface creep as given in the table 5.4. Transport of soil particles by saltation and surface creep is higher in heavy clay than loam.

Table 5.4. Transport of soil particles by suspension, saltation and surface creep in wind erosion process.

S. No.	Soil type	Soil moved by		
		Saltation (%)	Surface creep (%)	Suspension (%)
1.	Heavy clay	71.9	24.9	3.2
2.	Loam	54.5	7.4	38.1
3.	Fine sandy loam	54.7	12.7	32.6
4.	Sand	67.7	15.7	16.6

Source: Chepil (1945)

5.6.1 Factors affecting wind erosion

Soil clods prevent wind erosion due to large size. Coarse textured sandy loam, loamy sand and sands are more susceptible to wind erosion. These soils do not form stable clods due to low clay content and organic matter. Medium to fine textured loam, silt loam and silt clay loam soils form aggregation and can resist to the abrasive action of wind.

Rough surface formed by tillage operation reduces wind velocity and can trap saltating particles. A smooth soil surface is more erodible than a rough surface.

Wind erosion decreases with increasing soil moisture content. The dry soil particles are readily moved by wind. Moisture film holds the soil particles together and increases its stability against lifting by wind.

Fields with cover crops, residue mulch or shelter barriers reduce wind erosion as compared to bare field. The soil loss due to wind erosion decreases with increasing the density of cover crop.

Wind Erosion Control Measures

The basic principles in controlling wind erosion are:

i) Decreasing wind velocity at the ground surface

ii) Increasing the size of soil aggregates

iii) Keeping the soil moist

iv) Trapping the saltating soil particles

v) Covering the land surface

In arid and semiarid regions, the velocity of wind increases near the ground surface due to atmospheric turbulence on bare lands. The wind velocity can be reduced by growing crops or covering the land surface with residue mulch. Such covers acts as a barrier between soil and wind and effectively reduce the velocity of wind.

Repeated tillage operations decrease the stability of soil aggregates. Application of organic matter forms stable aggregates. Micro-organism binds the soil particles by producing a number of cementing substances.

Saltating soil particles can be trapped effectively by a vegetative cover or stubble mulch. Tillage operations create rough, cloddy surface which decrease the force of wind at the ground surface and trap moving particles.

Wind erosion decreases with increasing the moisture content of soil because water effectively holds soil particles by surface tension force. Soil wetness can be maintained by covering the land surface with crop, stubble mulch and land surface compaction.

The destructive action of wind can be reduced by providing shelter belts. The shelter belt consists of shrubs and trees spp. like Acacia, Eucalyptus, Ficus, Prosopis etc. are suitable for arid regions.

References

Chepil, W.S. (1945) Dynamics of wind erosion. Soil Science.60: 475-480

Narain, P., Chaudhary, R.S. and Singh, R.K. (1994) Efficacy of conservation measures in North – Eastern hilly regions. Indian Journal of Soil Conservation, 22: 42-62.

Olson, T.C. (1961). Determining K values. Report SCS, USDA 4-15743, 7-6: 25-28

Smith, D.D. and Wischmeir, W.H.C. (1962) Rainfall erosion. In Adv. In. Agron. 14: 109-148. Academic Press. New York.

Tejwani, K.G. and Bhardwaj, S.P. (1982). In Review of Soil Research in India. Part – II pp.608-621. 12th Int. Congress of Soil Sci., New Delhi, India, Feb. 16,1982

Study Questions

1. Write short notes on

i) Splash erosion

ii) Grassed water ways

iii) Surface creep

iv) Rainfall erosion index

v) Time of concentration

vi) Wind breakers and shelter belts

2. Justify the following statements with scientific reasons

i) Soil loss due to water erosion is severe in rolling topography.

ii) Soil texture is a static property of soil but, changes due to water erosion.

iii) A saturated soil encourages stronger splashing.

iv) Red and black soils are very prone to sheet erosion.

v) Siltation of finer particles are seen in delta region.

vi) Grasses and leguminous crops encourage infiltration capacity of soil.

vii) Cover crops reduces splash erosion.

viii) Contour bunds reduces the length of slope.

ix) Dry soils are prone to wind erosion.

3. What are the basic principles proposed in controlling wind erosion?

4. Differentiate between mechanical and biological methods of water erosion control measures?

5. Briefly discuss the role of grasses and legumes in soil conservation.
6. What are the basic principles proposed for soil conservation by water erosion?
7. How can we reduce the velocity of run off in rolling topography?
8. How does soil erosion plays a major role in degradation of our natural resources?
9. What biological control measures will you recommend for checking water erosion in agricultural lands?
10. How will you prevent and control soil erosion by wind?

6

Chemical Degradation Leading to Soil Acidity and Its Amelioration

Acid soils occupy about 30 percent of the land area of the world out of which 67% acid soils support forests and wood lands and 18 percent are covered by grassland vegetation. Only 4.2 per cent (179 million hectare) of the acid soils across the globe is used for arable crops.

Soil acidity is a major constraint in crop production. The removal of basic cations especially Ca and Mg by leaching and their replacement by acidic cations, H, Al and Fe on exchange complex leads to soil acidity. Soil developed from granite parent material are more acidic than those developed from basalt, shale or lime stone. In acid soils, low pH leading to toxic concentration of H^+, Al^{3+}, Mn^{2+} influences nutrient availability, organic matter decomposition and microbial activity.

Efficient soil amelioration and nutrient management practices are necessary for crops susceptible to soil acidity.

Distribution of acid soils

In India, about 90 M ha of the total geographical area are occupied by acid soils. Out of this, about 12 percent soils are strongly acidic (pH <5.0), 48 percent moderately acidic (pH 5.0 – 5.5) and 40 per cent mildly acidic (pH 5.6 – 6.5) in reaction.

Acid soils in India occur in the Himalayan region, the Eastern and North-Eastern plains, peninsular India and coastal plains under varying topography, geology, climate and vegetation. These soils are mostly distributed in the states of Assam, Manipur, Tripura, Meghalaya, Mizoram, Nagaland, Sikkim, Arunachal Pradesh, West Bengal, Jharkhand, Odisha, Chhattisgarh, Madhya Pradesh, Himachal Pradesh, Andhra Pradesh, Karnataka, Kerala, Maharashtra and Tamil Nadu.

Acid soils of Himalayan region are occupied by acid Podzols in association with brown forest soils. Acid soils of alluvial origin occur in West Bengal, Bihar, Assam and Odisha. Acid soils in marshy areas are found in Assam, Kerala, West Bengal, coastal districts of Odisha, south-east coast of Tamil Nadu, Terai regions of Uttar Pradesh, Bihar and West Bengal.

Acid sulphate soils are found in Sunderbans areas of West Bengal and Kuttanad area of Kerala.

The extent of occurrence of acid soils in different states of India is given in table 6.1.

Table 6.1. Acid soil regions (ASR) in India.

States	Area under ASR (M ha)	Percentage of ASR	Cultivated area with pH <5.5 (M ha)
Assam and North-Eastern States	20.0	80	3.5
West Bengal	3.5	40	2.0
Bihar	5.2	33	0.4
Odisha	12.5	80	1.8
Madhya Pradesh*	8.9	20	-
Andhra Pradesh	5.5	20	-
Tamil Nadu	2.6	20	-
Karnataka	9.6	50	-
Kerala	3.5	90	2.1
Maharashtra	3.1	10	0.3
Uttar Pradesh	2.9	10	-
Himachal Pradesh	5.0	90	0.1
Jammu & Kashmir	15.5	70	-

* Bihar including Jharkhand and Madhya Pradesh including Chhattisgarh *Source:*Mandal (1997)

Pedogenesis of acid soils of India

The factors which dominate in development of acid soils are rainfall, temperature, hydrological conditions and vegetation.

The major pedogenic process involved in the development of acid soils are:

i) Laterization

ii) Podzolization

iii) Intense leaching of bases

iv) Marshy conditions with significant amounts of undecomposed or partially decomposed organic matters.

Laterization

Laterites and lateritic soils are developed through the process of laterization.

The term laterite is derived from Latin Word **later** means ‘bricks’. The brick red color is due to the accumulation of Fe_2O_3 in various degrees of hydration.

Laterization is a pedogenic process in which silica is removed from the surface layer leaving sesquioxides (oxides of Fe and Al) to concentrate in the solum under wet and dry condition. Depending on the degree of laterization, the soils are termed as strong, medium or weakly laterite. The degree of laterization is determined on the basis of SiO_2: Al_2O_3 ratio.

If the ratio is less than 1, the soil is called laterite (Oxisols) and if the ratio is between 1-2, then it is called lateritic (Ultisols).

Humid (rainfall 1800-2500 mm) and tropical temperate (+30^{0}C) climate favour intense leaching, weathering of rocks and thus the soluble products of weathering materials are being dissolved and leached to deeper horizons.

The basic cations namely Ca^{2+}, Mg^{2+}, Na^+ and K^+ are quickly released during weathering of iron rich ferro-magnesium minerals namely biotite, chlorite, amphiboles, pyroxenes etc. and also rapid oxidation of organic matter. Under basic environment in absence of organic acid Fe, Al and Mn get precipitated and accumulated under high pH. Simultaneously the silica liberated from the parent materials is solubilized and leached leaving sesquioxides. As the alkaline bases are removed the residual soil is acidic in reaction. The laterites or lateritic soils are formed in B-horizon extending several meters below the surface layer. Laterite soils are usually granular, well drained, aerated, low in CEC and fertility. Phosphate fixation, soil crusting, toxicity of Fe and Al are common problems in laterite soils.

Ferruginous red soils

Red loam or latosols are developed through a milder process of laterization with evidence of varying degrees of podzolization having higher content of silica and lack of a lateritic layer in the upper horizon. The red soils are derived from ancient crystalline and metamorphic rocks under intense weathering. The alternate wet and dry periods coupled with high temperature enhanced the rate of weathering and oxidation of organic matter resulted in removal of soluble basic constituents like calcium, colloidal silica and simultaneous accumulation of secondary iron and aluminium oxides.

Podzolization

The term podzolization is derived from Russian Word '**Podzol**', meaning woodash like. In podzolization process, the sesquioxides are translocated in soil profile leaving behind SiO_2 resulting in formation of *Podzols* or *Podzolic* soils.

Under cold and humid climate at higher altitude with acid producing vegetation(coniferous pines), a dark colored substance is produced due to slow decomposition of leaves. The pH of soils fall below 5 in absence of bases.

Under acidic conditions and with limited microbial activity the organic acids react with sesquioxides and secondary clay minerals to form organic clay complex as well as organic sesquioxide complex which being soluble and leached to lower horizons with percolating water. During the process of hydrolysis of aluminium, the soil solution become very acidic and soluble iron and aluminium leach out from the surface horizon leaving SiO_2 giving grey or ashy appearance. As a consequence the pH of the A_2 horizon is the lowest in the soil profile and poorest in bases. The percolated solutes and silt and clay fractions get deposited in B_2 horizon and forms hard pan when dry.

The typical podzol soils showing distinctly acid leached ashy A_2 and characteristic B-horizon with pronounced deposition of humus and sesquioxides, in particular iron oxides, are lacking in India, but the effects of weak podzolization are usually noticeable. In India, the podzol soils are seen under deodar, blue pine, silverfir in hill forests of Kashmir Valley, coniferous podzol in Kullu, Kumanon hill soils in Almora, Garhwal and Nainital districts of Utter Pradesh.

Peat and marshy soils

Peat and organic soils are developed as a result of accumulation of appreciable amounts of organic matter in states of Assam, Kerala, coastal tract of Odisha and in certain regions of Bihar, Uttar Pradesh and in South –East coast of Tamil Nadu. In Assam, Surma Valley soils are characterized by their swampy nature and have peaty soils.

The peaty saline soils (Kari) in Kuttanad area of Kerala are black, highly acidic (pH <3.9) and contain 10-40 per cent organic matter, 20 to 30 per cent iron and aluminium but, poor in available phosphorus. Sometimes the soils have toxic concentration of iron and aluminium.

Source of hydrogen ions in soils

Soil acidity is measured using the pH scale. The pH scale expresses the concentration or activity of H^+ ions present in soil. In order to understand soil acidification and its management, it is necessary to know the sources and the processes of entry of H^+ ions into soils. Different sources of H^+ ions in soils are discussed below.

i) Carbonic acid

Root respiration and decomposition of organic matter by micro-organisms produce high level of CO_2 in soil. Formation of carbonic acid and its dissociation releases H^+ ion leading to soil acidity as per the following reaction:

$$CO_2 + H_2O \rightarrow H_2CO_3 \leftrightarrows HCO_3^- + H^+, pka = 6.35$$

Since, H_2CO_3 is a weak acid with relatively high pKa value, its contribution of H+ ions is negligible when the pH is below 5.

* pKa is the negative logarithm of the equilibrium constant for a dissociation reaction. For example;

$Ka = [H^+]\ x\ [H\ CO_3^-]\ /\ H_2CO_3$

Usually the dissociation of acid be considered negligible when the pH is 1 to 2 units lower than pka.

ii) Acid from biological metabolism

Production of H^+ion from dissociation of strong carboxylic or phenolic acid groups in humic substances is given by the reaction:

$RCOOH \leftrightarrows R\ COO^- + H^+ pka = 3\ to\ 5$

The pka of carboxylic groups ranged between3 to 5 depending on associated organic structures (R).

iii) Role of organic matter

The accumulation of organic matter tends to acidify the soil by two means. First organic matter associated with cations such as Ca^{2+} and Mg^{2+} forms soluble complexes, thus facilitate loss of these cations by leaching. Second, the organic matter is a source of H^+ ions because it contains numerous acid functional groups. On dissociation of H^+ ions, there is increase in number of negatively charged sites (pH dependent negative charge) as per the following reaction:

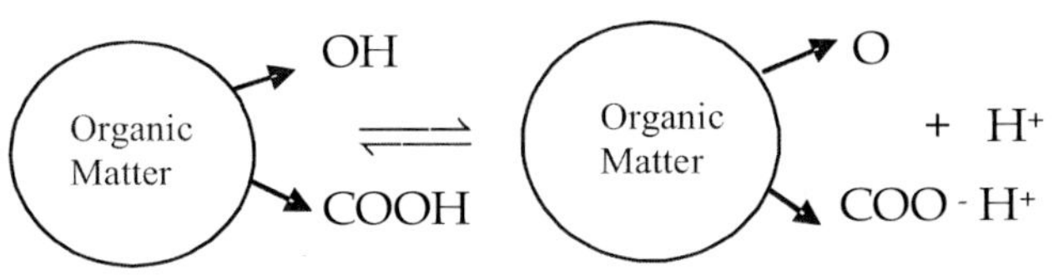

iv) Oxidation of nitrogen

Oxidation of basic ions to acidic ions is carried out by specific soil bacteria under aerobic condition. The oxidation process generally produce H^+ ions as per the reaction:

$$NH_4^+ + 2O_2 \rightleftarrows H_2O + 2H^+ + NO_3^-$$

Oxidation of Sulphur

Oxidation of pyrite (Fe S_2) in acid sulphate soils produced sulphuric acid (H_2SO_4) in presence of Thiobacillus spp. The reaction occurs as per the following:

$FeS_2 + 7/2\ O_2 + H_2O \leftrightarrows FeSO_4 + 2\ H^+ + SO_4^{2-}$

Chemical oxidation of pyrite is quite slow. But the rate of oxidation increases in presence of Genus Thiobacillus. These autotrops survive below pH 2.0 and derive energy from oxidation of reduced sulphur.

v) Aluminosilicate minerals

At low pH, most of the aluminium is present as the hydrated aluminium ions (Al^{3+}) which undergoes hydrolysis and release H+ ions in the soil solution:

$Al^{3+} + H_2O \leftrightarrows$ Hydrolysis $Al(OH)^{2+} + H^+$ (pH below 4.7)

$Al(OH)^{2+} + H_2O \leftrightarrows Al(OH)^+_2 + H^+$ (pH - 4.7 to 6.5)

$Al(OH)^+_2 + H_2O \leftrightarrows Al(OH)^0_3 + H^+$(pH-6.5 to 8.0)

$Al(OH)^0_3 + H_2O \leftrightarrows Al(OH)^-_4 + H^+$ (pH above 8.0)

It is also possible that structural OH^- ions at corner and edges may dissociate H+ ions and develop soil acidity.

vi) Plant Uptake cations

Plants maintain a balance between the positive and negative charges on the ions they take up from soil solution. Plants usually exude H+ ions into the soil solution when they take up certain cations or anions to maintain charge balance. The following example explain the exchange of ions between roots and soil solution.

Root		Soil solution
Example 1	← NH+	Uptake of cations
	→ H+	balance by release of
Example 2	← Ca^{2+}	H^+ ions from root. Decrease in pH
	→ $2H^+$	
Example 3	← Ca^{2+}	Uptake of cations
	← SO_4^{2-}	balance by uptake of anions.
		No pH change

Factors influencing acid soil formation

In certain situations, in addition to natural processes, the soil acidification is greatly accelerated by human activities.

In the areas of high rainfall, the bases like calcium and magnesium leached below the root zone leaving oxides of iron and aluminium. Countries having annual rainfall of more than 1000 mm per year have a high percentage of acid soils.

The soils developed from granite or any other acid parent materials are acidic in nature.

Use of ammonium based N fertilizers such as ammonium sulphate and ammonium chloride are oxidized in soil by microbes to produce strong inorganic acids by following reaction:

$(NH_4)_2\ SO_4 + 4O_2 \leftrightarrows 2HNO_3 + H_2SO_4 + 2H_2O$

The strong acids provide H+ ions that lower soil pH.

Equivalent acidity of N- fertilizers

The **equivalent acidity** refers to the amount of $CaCO_3$ required to neutralize the acidity developed by application of 100 kg ofN fertilizer. The acidity developed by application of 100 kg of ammonium sulphate requires 109 kg of $CaCO_3$ to get neutralized. The equivalent acidity of different N fertilizers are:

Urea - 74

Ammonium sulphate-109

Ammonium sulphate nitrate-93

Ammonium chloride-128

Calcium ammonium nitrate-neutral

Addition of organic manures or sewage sludge on agricultural and forest lands can decrease soil pH, both by oxidation of nitrogen released and by formation of organic and inorganic acids during decomposition. Humus react with Fe and Al ions form complexes which may subsequently undergo hydrolysis to yield hydrogen ions.

Acid rain is another cause of soil acidification. Industrial activities such as combustion of coal in electric power stations, combustion of fuel in vehicles and smelting of sulphur containing metals emits enormous quantities of sulphur and nitrogen containing gases into the atmosphere. The gases mainly sulphur dioxide and oxides of nitrogen react with water in the atmosphere form HNO_3 and H_2SO_4. Carbon dioxide released through burning of crop residues, decomposition of organic matter forms carbonic acid. These strong acids return to earth in rain is called **acid rain** since, its pH is one or two units lower than un-polluted rain. The pH of acid rain commonly varies between 4 to5 but, may be as low as 2.The greatest amount of acidity falls on land makes it acidic.

Pools of soil acidity

The major pools of acidity common in soils are:

i) Active acidity

ii) Exchange acidity

iii) Residual acidity

iv) Potential acidity

i) Active acidity

The H+ ions in soil solution contribute active acidity which is measured in terms of soil pH. This pool is very small compared to exchange and residual acidity. About 2 kg of $CaCO_3$ is needed to neutralize active acidity up to 15 cm upper soil layer of a hectare of land. However, the active acidity is very important, since it determines the solubility of many substances and provides an environment to which plant roots and microbes are exposed.

ii) Exchange acidity

The H+ and Al^{3+} ions present in exchange complex of soils contribute exchange acidity. H^+ and Al^{3+} concentration in soil solution (active acidity) and exchange complex refers to exchangeable acidity. It is the acidity developed due to adsorbed H^+ and Al^{3+} ions on the soil colloids, which are in equilibrium with soil solution. These ions can be released into soil solution by cation exchange when treated with an unbuffered salt, such as KCl.

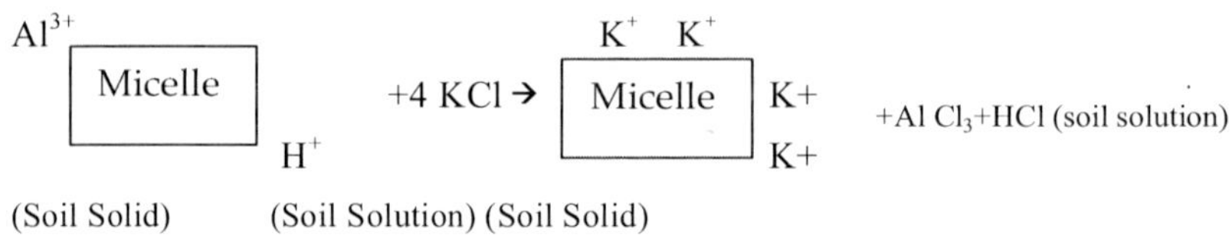

Exchangeable acidity in strongly acid soils is commonly thousand times higher than that of active acidity. The amount of $CaCO_3$ required to neutralize this acidity is about 100 times more than that needed to neutralize active acidity.

iii) Residual acidity

The H^+ and Al^{3+} ions associated with organic matter and clays and subsequent release and precipitated as $Al(OH)_3$ contribute to the pool of residual acidity. The residual acidity remained in soil after active and exchange acidity has been neutralized.

Residual acidity = Total acidity - exchange acidity - active acidity

The release of H^+ and Al^{3+} ions from the organic complex with liming (i.e.

$Ca(OH)_2$) is stated as :

$$\boxed{\text{Micelle}}\begin{matrix}Al^{3+}\\ \\ H^{+}\end{matrix} + 2\,Ca(OH)_2 \leftrightarrows \boxed{\text{Micelle}}\begin{matrix}Ca^{2+}\\ \\ Ca^{2+}\end{matrix} + Al(OH)_3 + H_2O$$

(Soil Solid) (Soil Solution) (Soil Solid) (Soil solution)

The residual acidity is far greater than either active or exchange acidity. The amount of ground lime stone required to partly neutralize the residual acidity in 15 cm upper soil is about 5 to 10 metric tonnes per hectare.

iv) Potential Acidity

Potential acidity is titratable but non exchangeable. Aluminium hydroxy ions and H^+ and Al^{3+} ions present in non- exchangeable forms with clays and organic matter accounts for the potential or reserve acidity of a soil. In acid sulphate soils, potential acidity developed due to oxidation of pyrite (FeS_2) is given as:

$$2FeS_2 + 7\tfrac{1}{2}\,O_2 + 5\,H_2O \leftrightarrows 2Fe\,OOH + 8H^+ + 4\,SO_4^{2-}$$

The potential acidity of sulfide – rich soils can require greater amounts of lime stone for neutralization than that required for residual acidity. The amount of potential acidity is difficult to quantify, because the rate of its reaction with bases is generally slower than the exchangeable acidity.

Per cent base saturation

The per cent base saturation is the ratio of the exchangeable bases to the cation exchange capacity of soil as:

$$\text{Per cent base saturation} = \frac{\text{Exchangeable } Ca2^+ + Mg2^+ + Na^+ + K^+ [c\ mol(+)\ Kg^{-1}]}{\text{Cation exchange capacity } [c\ mol(+)\ Kg^{-1}]} \times 100$$

Broadly, low per cent base saturation ie. below 50 per cent and a pH around 5.0 indicates soil acidity. When the base saturation increases to 80 per cent, the pH is 6.5, an almost neutral soil exists.

Buffering of soils

Buffering refers to resistance to sudden change in pH. When 1 ml of 0.01 N HCl is added to 1 lit of pure distilled water of neutral pH, the resulting solution would have a pH of about 5.0. If the same amount of acid is added to a soil suspension, the change in pH would be very small. Soils behaves like a buffered weak acid and that will resist sharp changes in pH.

The buffering mechanism may be explained as follow

Residual acidity		Exchangeable Acidity		Active acidity
Al and H bound on clay and humus	⇆	Al^{3+}, $Al(OH)^{y-}_{x}$ and H^{+} ions held in exchange site	⇆	Al3+, $Al(OH)^{y-}x$ and H^{+} ions in soil solution

In soils with intermediate pH levels (5 to 7), buffering can be explained in terms of the equilibrium that exists between the three principal pools of soil acidity: active, exchangeable and residual acidity. If enough base (lime) is applied to neutralize H+ ions in the soil solution, they are largely replenished as the reaction move to right, thereby minimizing the change in soil solution pH. On the other hand, if the H+ ion concentration of soil solution is increased (by organic decay or fertilization) the reactions are forced to the left consuming H+ ion and again minimizing changes in soil solution pH. Because of the involvement of residual and exchangeable acidity, soils with higher clay and organic matter contents are likely to be better buffered in this pH range.

Factors affecting buffering capacity of soils

i) Soils having more of clay and organic matter exhibit higher buffering capacity, resisting the change in soil pH.

ii) Other influencing factors remaining constant, higher the exchange capacity of a soil the greater is its buffering capacity.

iii) The degree of buffering is highest between pH 4.5 to 6.0 and drops off below pH 4.5 and above 6.0.

iv) Soils containing large amounts of calcium carbonate, organic acids and phosphates show higher buffering capacity.

v) Soils containing large amounts of 1:1 type clays are less buffered than the soils dominated with 2:1 type of clay minerals.

Importance of soil buffering capacity

The importance of the buffering capacity of soils is of two folds:

i) Stabilization of soil pH: Buffering tends to ensure in stability of soil pH, preventing drastic fluctuations that might be detrimental to plants, soils, microorganisms and aquatic ecosystems.

ii) Buffering influences the amount of amendments such as lime or sulphur, required to bring a change in soil pH to a desired level.

Role of aluminium in soil acidity

The Al^{3+} ions released from clay minerals by cations are hydrolyzed to monomeric and polymeric hydroxyl aluminium complexes. At low pH values, most of the aluminium is present as hexahydrated Al $(H_2O)_6^{3+}$ ion.

Hydrolysis of hexa hydrated aluminium by the stepwise reaction with liberation of hydronium ion (H_3O^+) is stated as:

$$Al(H_2O)_6^{3+} + H_2O \leftrightarrows Al(OH)(H_2O)_5^{2+} + H_3O^+$$

$$Al(OH)(H_2O)_5^{2+} + H_2O \leftrightarrows Al(OH)_2(H_2O)_4^{+} + H_3O^+$$

$$Al(OH)_2(H_2O)_4^{+} + H_2O \leftrightarrows Al(OH)_3(H_2O)_3^{0} + H_3O^+$$

$$Al(OH)_3(H_2O)_3^{0} + H_2O \leftrightarrows Al(OH)_4(H_2O)_2^{-} + H_3O^+$$

Each successive step occurs at a higher pH.

Total aluminium in soil solution is pH dependent and include all above hydrolysis species. These aluminium products can be readsorbed on surface of clay minerals and organic matter causing further hydrolysis with the release of H+ ions into soil solution and thereby develops soil acidity.

Aluminium toxicity

Aluminium toxicity is a major growth limiting factor for plants in strongly acidic soils with pH < 5.0 to 5.5. Aluminium toxicity limits the microbial breakdown of organic matter, interfere with cell division in root tips and lateral roots, decrease root respiration, fix soil P in less available forms and interfere with the uptake, transport and use of several plant nutrients i.e. Ca, Mg, K, P, Fe and Mo. In most plants, aluminium taken into the root remain there, little is translocated to the shoot. But in aluminium accumulator plants such as tea, much of alluminium is taken up and accumulated in leaves (around 5000 ppm aluminium).

Soil reaction

Soil reaction is an important property of the soil that influences physical, chemical and biological behavior of soil. Plant nutrients transformation, availability, uptake by crops and microbial activities determine the production potentiality of soils. The concentration of hydrogen and hydroxyl ions of a solution determine acidity and alkalinity.

Water molecules ionize to hydrogen and hydroxyl ions. In pure water, the dissociated ions remain in equilibrium with un-dissociated water molecules.

$$H_2O + H_2O \leftrightarrows H_3O^+ + OH^-$$

This equation reveals that the hydrogen ions in aqueous solution are not H+ but, contain at least one water molecule forming hydronium ions, H3O+.

As per the **law of Mass** action, the equation leads to

$$Kw = \frac{aH_3O^+ \times a\,OH^-}{aH_2O \times aH_2O}$$

However,

$Kw = aH_3O^+$ x a OH^- Or, Kw = aH+ x a OH^- where,

Kw- is the ionic product of water, a_H^+ and a_{OH}^- -are the activity of H+ and OH^- ion, respectively. It is difficult to express such activities in conventional chemical forms.

Danish scientist **Sorensen**, 1909 first proposed a logarithmic scale to denote pH scale ranging from 0 to 14 with 7 being neutral.

The H+ ion concentration is usually small and expressed in terms of pH.**The pH is defined as negative logarithm of H+ ion activity expressed in moles per litre,**

Where, **activity(a) =concentration (c) x activity coefficient (f) symbolically the pH is given by,** $pH = 1 / \log aH^+$

The pH of the scale is derived from the dissociation constant of water, Kw.

Water is dissociated into hydrogen ions and hydroxyl ions as:

$$H_2O \leftrightarrows H^+ + OH^-$$

The extent to which water ionizes can be expressed in terms of an ionization constant, Kw.

$$Kw = [H+]\,[OH^-] = 10^{-14} \text{ (at } 25^0C)$$

$[H^+]$ and $[OH^-]$ are the concentration of hydrogen and hydroxyl ions expressed as moles per litre.

The value of Kw is 10^{-14} at 25^0C and in any aqueous solution at this temperature, the products of the concentration of H^+ and OH^+ ions is 10^{-14}.

Dividing both sides of above equation by 1'and taking logarithms, the equation is stated as log 1/Kw = log 1/ [H+] + log 1/ [OH-] = 14

or, pH + p OH = 14

Therefore, an inverse relationship exists between pH and pOH. If the pH of a solution is 6, then pOH of that solution is 8.

Example

At pH 6.0, the a_{H+} is calculate as, pH = - log aH+ = 6.0 or, log a_{H+} = -6 = log 10^{-6} or, a_{H+} = 10^{-6} moles per litre. Similarly, aOH^- = 10^{-14} / 10^{-6} = 10^{-8} moles per litre.

In pure water at 25°C, pH + p OH = 14 and pH = p OH = 7.0

At 25°C, a pH value less than 7 indicates an acidic solution and that more than 7, an alkaline condition. The pH 7 represents a neutral solution.

The relationship between pH, pOH, H+ ion and OH^- ion concentration is given in table 6.2.

Table 6.2. The relationship between pH, pOH, H+ ion concentration and OH^- ion concentration.

pH	pOH	H+ ion conc. (Moles L^{-1})	OH- ion conc. (Moles L^{-1})
0	14	10^{-0}	10^{-14}
1	13	10^{-1}	10^{-13}
2	12	10^{-2}	10^{-12}
3	11	10^{-3}	10^{-11}
4	10	10^{-4}	10^{-10}
5	9	10^{-5}	10^{-9}
6	8	10^{-6}	10^{-8}
7	7	10^{-7}	10^{-7}
8	6	10^{-8}	10^{-6}
9	5	10^{-9}	10^{-5}
10	4	10^{-10}	10^{-4}
11	3	10^{-11}	10^{-3}
12	2	10^{-12}	10^{-2}
13	1	10^{-13}	10^{-1}
14	0	10^{-14}	10^{-0}

Factors affecting pH measurement

Several factors affect the pH measurement of a particular soil sample. The factors are:

i) Soil : Water ratio

As the soil suspension becomes more diluted the pH of an acid soil rises. This phenomenon is called '**Suspension effect**'. Several reasons have been given for this effect. It is assumed that in acid soils, the concentration of H^+ or Al^{3+} ions is greatest near the negatively charged colloidal particles and a gradient exists from the particles into the soil solution. Therefore, the thicker the paste into which the glass electrode is placed, the more is the contact of the electrode in the compressed exchange acidity surrounding the colloidal particles and lower the pH value. As more water is added, the ion layer would be more diffuse and the electrode would measure the pH of the liquid phase of the suspension.

Another reason may be the effect of '**Junction Potential**" present in clay – water systems. The potentiometric measurement of pH requires the transference of both K^+ and Cl^- ions of the salt bridge at same rate. In clay systems the K+ ion would be attracted to the negatively charged particles and therefore, would not diffuse in same rate as the Cl^- ion. This phenomenon would give rise to a potential difference at the junction of salt bridge in the clay system. The thicker is the paste, the greater is the potential difference and would lead to error in pH measurement.

This problem can be elevated by measuring soil pH in a wide soil : water ratio by placing the calomel half–cell in supernatant liquid and glass electrode in the sediment.

ii) Soluble salts

When the concentration of salts in soil-water suspension increases the pH decreases. The salts will undergo cation exchange with acidity in soil to release acid to soil solution resulting decrease in pH. This problem can be eliminated by using 0.1N KCl or 0.01 M $CaCl_2$ solution instead of water. It is assumed that the salt concentration in soil sample is negligible with respect to the amount of salt added into the solution.

Some researchers used "**Lime Potential**" instead of pH of soil because the value of lime potential for a particular soil is constant over a fairly wide range in conditions.

The "**Lime Potential'** is calculated as:

pH - ½ p(Ca+Mg) where p repents the negative logarithm of activity of Ca+Mg.

iii) Presence of electrolytes

Presence of electrolytes also decrease soil pH in neutral or alkaline range. When a Na-clay is dispersed in pure water, a small portion of Na ions will hydrolyze from the diffuse layer. Addition of an electrolyte suppressed the hydrolysis of Na resulting in decrease in pH.

At an equal concentration, $CaCl_2$ lowers the pH of a soil more than does NaCl. Fluctuations in electrolyte concentration in soil solution is influenced by temperature, oxidation of organic matter, leaching action of rain or irrigation water and growing season.

iv) Carbon dioxide

The pH of soil containing calcium carbonate, carbonate or bicarbonate in soil solution is closely related with carbon dioxide pressure, concentration of bicarbonate ion and ionic strength of the soil solution. The relationship is given by:

$$pH = pK_1 - 0.5\sqrt{M+} \times \log\frac{(HCO_3)}{(CO_2)} \text{ where,}$$

pK_1 – negative logarithm of first dissociation constant of carbonic acid (6.26 at 24^{0}C), M- Ionic strength of the soil solution and (CO_2) – concentration of CO_2 in moles per litre. The pH of calcareous soils is reduced in proportion to the logarithm of the CO_2 pressure of the soil air.

Soil pH and nutrient availability

The availability of several plant nutrients is influenced by soil pH. In strongly acid soils, the availability of macro and secondary nutrients (N, P, K, Ca, Mg and S) as well as molybdenum and boron is decreased. In contrast, the availability of most micronutrients such as Fe, Mn, Zn and Cu increased by low pH, even to the extent of toxicity to higher plants.

In slightly acidic to moderately alkaline pH range, the availability of all macro and secondary nutrients are plentily available but the levels of available Fe, Mn, Zn and Cu are so low that the plant growth is affected. The availability of boron increases within pH range of 5.5 to 7.5 and Mo between pH 6.0 to 9.0. Maximum P availability occurs at a pH ranges of 5.5 to 7.5.

It is difficult to analyze the relationship between pH and nutrient availability. However, the pH range of 6.5 to 7.5 may provide satisfactory plant nutrient levels overall.

Optimum pH for field crops

Majority of the agricultural lands have soil reaction neither too much of acidic or alkaline to cause direct damage to crop roots. However, the roots of various crops require an optimum pH range for best growth. When the pH is not within range the roots and the crop as a whole suppressed. The optimum pH range for some crops are listed in table 6.3.

Table 6.3. Optimum pH range for some crops.

Crops	pH range
1. Field crops	
Maize and wheat	5.5-7.5
Rice	5.0-6.5
Groundnut, greengram, black gram, gram, cowpea, guanr, tobacco, mustard	5.5-7.0
Peas, bean, soyabean	6.0-7.5
Cotton	5.5-6.5
Sugarcane	6.0-8.0
2. Vegetable crops	
Brinjal	5.5-6.0
Onion, carrot, cucumber	5.5-7.0
Tomato, cauliflower	5.5-7.5
Beet, cabbage, okra	6.0-7.5
Radish	5.0-7.0
Potato	4.8-5.6
3. Fruit crops	
Coconut, papaya	6.0-8.0
Banana	6.0-7.5
Citrus	6.0-8.5
Mango, jackfruit	5.0-7.0

Soil reaction and crop production constraints

The crop production constraints at different pH range is summarized and given below:

Soil pH range	Production constraints
<5.0(Strongly acidic soils)	High P-fixation and low availability. Calcium and Mg deficiency in crops like cotton, groundnut, okra, citrus etc. Molybdenum deficiency in cole crops and legumes. Boron deficiency in cole crops and vegetables. Iron toxicity in low land paddy and Al toxicity in uplands. Poor growth of legumes .
5.0 – 6.0 (Moderately acidic)	Problems as above
6.0- 7.5 (Slightly acidic to neutral)	No soil reaction problems
7.5-8.5 (Slightly to moderately alkaline)	Phosphate fixation and low availability. Probable K and Mg deficiency in calcareous soils. Deficiency of Zn, Fe, Mn, Cu and B in crops particularly in vegetables and fruit crops. Potato tubers are affected by scab disease.
Above 8.5 (strongly and very strongly alkaline)	Direct root damage, deficiency of Ca, Mg and S besides acute deficiency of Fe, Mn, Zn, Cu, Mn and Co. Boron toxicity in crops and Mo toxicity in forage crops.

Characteristics of acid soils

i) Physical properties of acid soils

Acid soils usually occur in high rainfall areas where leaching of bases, fine particles and organic matter is the predominant pedogenic processes. The upland and low land soils have distinct physical properties.

Upland acid soils

The upland acid soils are coarse textured, usually compact at soil surface and have low organic matter besides distinct soil acidity. The soil textures are usually loamy sand, sandy loam or loam. The structure of soil is very weak. High content of Fe and Al develops usually crumb or granular structure. Due to coarse soil texture, drainable porosity is high with low water and nutrient retention capacity. The soils have high permeability of air and water. Surface crust formation after a heavy rainfall followed by a dry spell is a peculiar characteristics of most upland soils due to presence of Fe and Al. The bulk density of these soils raise to as high as 1.8 g cm^{-3} and hampers germination of seeds and poor plant growth. Due to low water retention capacity, the soils become dry after 2 to 3 days of rainfall or irrigation.

Major acid soils regions in India are rainfed and the crops suffer due to water stress during dry spells. The **zero soil water flux** rapidly moves downward causing water deficit in surface layers. Thus, the shallow rooted cereals are more frequently affected during dry spells as compared to deep-rooted crops like pulses, arhar, castor etc.

The leaching loss of plant nutrients in upland soils are very high. The NH_2 – N and NO_3-N are leached below the root zone within 24 hours of its application, but NH_4-N is retained with 2-5 cm of soil surface for a longer period. The cations like Ca, Mg and K are leached down due to presence of 1:1 type clay minerals.

Low land acid soils

The acid soils present in low lying areas are of two types. Soils present in coastal areas (alluvial, coastal saline soils) are slightly acidic whereas, the valley bottom soils under undulating terrain are slightly acidic to neutral in reaction. Drainage is a major problem besides soil acidity in low land soils. The infiltration rate is quite low leading to water logging. The hydraulic conductivity of the soil is moderate to low causing frequent rise of ground water table and super saturation for prolonged period. Fine textured soils containing low organic matter promotes unfavorable soil structure. Angular blocky or prismatic soil structure developed in sub surface horizon. Water holding capacity of soil is high. When such soils are puddled for transplanted rice, a hard compact layer is developed having bulk density as high as 1.8 g cm^{-3} in sub surface layer. When such soils are dried during post moon season, surface cracking and clod formation become common features in these soils. Thus, fine texture, unfavorable soils structure, plow pan in wet season and development of clod in dry season are some of the soil constraints for crop production in lowland soils.

The crop production constraints with respect to physical properties are summarized as:

i) High percolation loss in coarse textured soil

ii) Crust formation in red soils

iii) Shallow soil depth causing quick drying of soils

iv) Compact layer in subsurface horizons

v) Drainage problem in deltaic coastal region and valley soils

vi) Crop water stress due to low water retentivity

vii) Erodibility in red soil areas

ii) Chemical properties of acid soils

Low pH

Acid soils in India have been formed due to drastic weathering associated with hot humid climate and heavy rainfall. Pedogenic processes like laterization, podzolization and humification contribute to soil acidity.

Based on the soil pH, the acid soils are classified into 6 groups as:

S.No.	Degree of soil acidity	pH range
1.	Extremely acidic	less than 4.5
2.	Very strongly acidic	4.5 – 5.0
3.	Strongly acidic	5.1-5.5
4.	Moderately acidic	5.6-6.0
5.	Slightly acidic	6.1-6.5
6.	Neutral	6.6-7.3

Based upon the **pedogenic processes** and soil properties, acid soils of India has been classified broadly into ten groups. The broad acid soil groups of India with pH are given in Table 6.4.

Table 6.4. Broad acid soils groups of India.

S.No.	Broad soil groups	pH range
1.	Laterites	4.8-7.0
2.	Lateritic red soils	5.0-7.0
3.	Ferruginous red soils	5.0-6.8
4.	Mixed red and yellow soils	5.5-6.5
5.	Podzolic and brown forest soils	5.1-5.4
6.	Foot hill soils	4.5-6.5
7.	Peat soils	3.5-4.5
8.	Acid sulphate soils	3.0-4.5
9.	Degraded saline or saline acid soils	5.2-6.6
10.	Alluvial acid soils	3.5-5.8

Source: Murthy *et al.* (1976)

Organic matter and nitrogen status

The organic matter and nitrogen content of acid soils in India showed extreme variations because of climate, vegetation and intensity of organic matter decomposition.

In general, laterites, lateritic and red loam soils are characterized by low status of organic matter and nitrogen, whereas hill and forest soils are comparatively high (Table 6.5).

Hill districts soils of Assam, alluvial soils of Brahmaputra valley, terai soils of West Bengal, podzol and brown forest soils of Utter Pradesh have higher content of organic matter and nitrogen.

Table 6.5. Organic matter and nitrogen status of some acid soils of India.

Soil	Organic Carbon (%)	Nitrogen (%)
Brahmaputra valley soils (Assam)	1.2-2.20	0.08-0.11
Old alluvial soils (Assam)	1.02	0.11
Laterite and Lateritic soils (Assam)	1.02	0.11
Hill soils (Assam)	1.26	0.18
Forest soils (Assam)	1.59	0.21
Laterite & lateritic soils (WB)	0.05-0.5	0.01-0.8
Red soil (WB)	0.05-0.5	0.01-0.05
Terai soils (WB)	0.08-3.0	0.09-0.02
Red soil (Kanke, Ranchi)	0.08-0.40	0.03-0.05
Red soil (Maharashtra)	0.48	0.05
Acid soil (HP)	0.7	0.08
Nilgiri soil (TN)	0.7	0.07
Brown forest soil (UP)	1.90	0.15
Podzol soil (UP)	3.30	0.25
Laterite soil (Bhubaneswar, Odisha)	0.36	0.05

Source: Mandal (1997)

Phosphorus status in acid soils of India

Most of the acid soils of India are poor in phosphorus. Phosphates are predominately in the form of iron phosphate in the acid soils of Bihar, Maharashtra, Odisha.

Phosphate fractions of Assam soils indicates that the Fe-bound phosphate is higher than Al-Phoshphates. Acid soils of alluvial origin contain higher proportion of organic-P than in-organic P.

It is reported that all soils of Meghalaya, Tripura, Manipur and Arunachal Pradesh and over 50% soils of Nagaland, 40% of Mizoram and 30% of Sikkim are rated as low in available P. In Odisha, red and lateritic soils are deficient in available P (1.3 to 5.9 ppm).

Problem of P – fixation

In strongly acid soils having pH less than 5.5, the concentration of Fe and Al ions in soil solution exceeds that of H_2PO_4 ions. Consequently the latter are precipitated as insoluble phosphates of Fe and Al. Aluminium and iron bound phosphates are relatively less available to crops. Similarly the hydrous oxide of Fe and Al, which are present in soil colloids of acid soils, adsorb water soluble H_2PO_4 ions and render them less available to crops. Most laterite and lateritic soils and some red soils have Kaolinite as the dominant mineral in their clay fractions. The reaction between the exposed OH^- groups on mineral surface and $H_2PO^-_4$ ions also results in fixation of applied P.

Critical limits of available soil P in acid soils

A better approach for prediction of crop response to applied P- fertilizers is based on the determinations of critical limits of P for individual crops. The critical limits of P for some crops are given in table 6.6.

Table 6.6. Critical limits of P (Olsen's method) for some crops in acid / neutral soils.

Crop	Soil type	pH	Critical limit (P_2O_5 Kg / ha)
High-yielding rice	Coastal alluvial (Kerala)	4.3-6.5	67.0
	Red soils (Karnataka)	4.9-7.1	15.0
	Coastal alluvial (TN)	5.4-7.5	69.0
	Laterites(Puri, Odisha)	5.2-7.4	71.0
	Deltaic alluvium (Cuttack, Odisha)	4.1-7.5	25.2
	Mixed red & lateritic (Sambalpur, Odisha)	5.0-7.4	57.3
	Lateritic soils (Konkan, Maharashtra)	4.9-6.7	46.2
Jute	Alluvial (WB)	5.5-7.3	25.0
Potato	Brown forest hills soils (Shimla)	5.4-6.3	82.1
	Brown Forest hill soils (Meghalaya)	4.7-5.5	106.2 (Bray's P1)

Rock phosphate-a source of P in acid soils

Phosphate fixation is a major problem in acid soils. In agriculture the cost of P fertilizers can be reduced by direct use of low grade rock phosphate alone or in combination with water soluble single super phosphate. In combined application (R.P: SSP in 1: 1 ratio), phosphorus requirement of crops at early stage could be met from water soluble P and P released from rock phosphate could satisfy the crop in subsequent period. Partial acidulation of rock phosphate has been observed to be possible means for efficient utilization of low grade rock phosphates. Udaipur rock phosphate containing substantial amount of calcite and dolomite served as an amendment when applied directly to acid soils.

Potassium status of acid soils of India

Potassium status of acid soils of India is characterized by difference in composition of parent materials and the degree of K-bearing minerals. Available data on total K content of acid sols indicate that laterite, lateritic and foot hill soils range between 0.1 to 0.4% K_2O. Acid soils of sedentary origin where there is abundance of mica and K-bearing feldspars are rich in potassium. The sand, silt fractions in soils contributed towards total K-content whereas clay fraction contributed towards non- exchangeable K.

Secondary and micronutrients in acid soils

Most of the acid soils are deficient in Ca, Mg and S except acid sulphate soils which contain high amount of sulphur. Base saturation in most of the acid soils is generally low. In soils having calcium saturation less than 25% of total CEC, most crops requires calcium application. In strongly acid soils, high saturation of Al (65-70%) is frequently associated to reduce absorption of Ca and Mg by plants. High application of N and K to soils also depress the level of Mg content in plant tissues.

Low soil pH in acid soils enhance the activity of Al, Fe and Mn oxides in soil solution and thus decreases the availability of several micronutrients to the growing plants.

The analysis of 68,680 soils samples collected from different states representing mostly acid soils indicated the deficiency of Zn, Cu, Fe and Mn by 31, 3, 4, and 3 per cent respectively, in comparison to 47, 4, 14 and 6 percent for non-acidic soils in India (Table 6.7). The deficiency of Zn was minimum of 8% in Pondicherry and that of maximum to 53% in Chhattisgarh due to low organic matter content. The deficiency or sufficiency levels of micronutrients in soils were categorized based on the critical concentrations as:

0.6 ppm Zn, 0.2 ppm Cu, 4.5 ppm Fe, 2 ppm Mn, 0.5 ppm hot water soluble B and 0.15 ppm ammonium oxalate extractable Mo.

The deficiency of B and Mo was 45 and 46%, respectively. Boron deficiency in acid soils may be attributed to a number of factors such as high rainfall, high acidity, coarse texture and low oxidation of organic matter.

The data presented in table 6.7 clearly demonstrated that most of the acid soils are adequate in Fe, Cu and Mn whereas deficiency of Zn and B have been reported in old alluvial soils of Assam, red and yellow soils of Odisha, red, laterite and alluvial soils of West Bengal.

Table 6.7. Extent of micronutrient deficiency in acid soils of India.

State	Percent deficiency*					
	Zn	Cu	Fe	Mn	B	Mo
Andhra Pradesh	15	1	2	1	52	53
Tamil Nadu	37	7	16	7	21	-
Karnataka	41	0.1	4	2.6	32	-
Kerala	34	3	0.9	-	-	78
Pondicherry	8	4	2	3	-	-
Assam	34	1	2	2	17	-
Jharkhand	42	3	1	1	57	55
Chhattisgarh	53	1	1	1	49	30
Odisha	19	2	1	1	69	87
West Bengal	36	0	0	3	68	-
Meghalaya	57	2	0	23	-	-
Maharashtra	30	1	2.4	1	-	-
Jammu & Kashmir	12	-	-	-	-	-
Himachal Pradesh	42	1	27	5	1	-
Uttaranchal	28	7	14	9	14	-
All India	31	3	4	3	45	41

* No of samples tested for Fe, Mn, Zn and Cu - 68,680, for B – 24619 and for Mo – 1180.
Source: Singh (2006).

Toxic effects of nutrients in acid soils

The concentration of iron and manganese in solution depends upon the soils pH, soil organic matter and intensity of soil reduction. Under reduced condition, concentration of the elements like Mn^{4+} and Fe^{3+} reduced to Mn^{2+} and Fe^{2+}, respectively and increases their concentration to toxic level under impeded drainage condition. The growth and yield of rice decreased drastically because of these toxic ions.

Toxicity of aluminium is a major problem in uplands under strongly acidic condition. Aluminium toxicity affects plant growth by restricting root growth, restricts the absorption and translocation of P, Ca, Fe and Mn. It also restricts microbial activity in soils.

Microbial Activity

Soil acidity affects not only plant growth but also biological activity in soils. Under acidic environment, certain groups of micro-organism are stimulated and other retards. Activities of particular groups of micro-organisms change with degree of soil acidity.

In acid soils of Kanke (Ranchi) (pH 5.4), the population of bacteria andactinomycetes were lower whereas the fungal populations were several times higher than in other soils.

The activity of nitrogen fixing bacteria such as non-symbiotic free living bacteria, non-symbiotic free living blue green algae and symbiotic bacteria is low in acid soils, but enhanced by phosphate, lime and molybdenum application. In addition to organic matter decomposition and nitrogen mineralization, several other biochemical processes such as nitrification (Nitrosomonas, Nitrobacter), denitrification (Thiobacillus denitrificans), phosphate solubilization (Bacillus, Pseudomonas), oxidation of iron (Siderobacter, Siderophacus) and manganese (Azotobacter, Chroococum) occurs\at different degree of soil acidity.

Amelioration of acid soils

Liming

Liming in agriculture refers to addition of any calcium or calcium and magnesium containing compounds to reduce soil acidity.

The total acidity of a solution is the sum of the concentration of the active acidity and potential acidity which is illustrated by:

$$\text{Clay micelle} \left.\begin{matrix} Ca^{2+} \\ H^+ \\ Al^{3+} \\ H^+ \\ K^+ \end{matrix}\right\} \rightleftharpoons \left\{\begin{matrix} Ca^{2+} \\ H^+ \\ Al^{3+} \\ \\ K^+ \end{matrix}\right. \text{Clay micelle} + H^+$$

The hydrogen and aluminum held to clay micelle represent the potential acidity and dissociated H^+ in right side of this expression represent the active acidity. When a base is added, the active hydrogen ions neutralized, the equilibrium will tend to readjust itself and some of the hydrogen ions held by clay will dissociate.

There will be only a slight change in active acidity of the system. This process can be continued, until all the potential hydrogen ion is neutralized and adsorbed aluminium is precipitated as aluminium hydroxide.As the neutralization process is continued, there is gradual decrease in active acidity with a concomitant increase soil pH. The amount of lime to be applied considers both active and potential acidity.

Liming reactions in acid soils

Most of the liming materials when applied to acid soil react with carbon dioxide and water to produce bi-carbonate. This reaction can be illustrated for $CaCO_3$ as:

$$CaCO_3 + H_2O + CO_2 \rightarrow Ca(HCO_3)_2$$

$$Ca(H\ CO_3)_2 \rightarrow Ca^{2+} + 2\ HCO_3^-$$

(calcium ion takes part in cation exchange)

$$HCO_3^- + H^+ \rightarrow H_2CO_3 \leftrightarrows CO_2 + H_2O$$

(from lime)(From solution)

In this reaction the HCO_3^- ion is capable of consuming H^+ ions (from active acidity) to form weak acid and the Ca^{2+} ion from lime stone takes part in ion exchange. The acidity of soil get neutralized and the base saturation of colloidal material is increased.

The carbonate liming materials also react directly with the exchangeable residual acidity in acid soils. The Ca^{2+} replace H+ and Al^{3+} on the colloidal complex as illustrated below:

$$\boxed{\text{Clay or humus}}\begin{matrix} H^+ \\ Al^{3+} \end{matrix} + 2CaCO_3 + H_2O \rightarrow \boxed{\text{Clay or humus}}\begin{matrix} Ca^{2+} \\ Ca^{2+} \end{matrix} + Al(OH)_3 + 2\ CO_2$$

The insolubility of Al $(OH)_3$, the weak dissociation of water and the release of CO_2 gas to atmosphere pull this reaction to the right. The adsorption of the calcium ions lower the percentage of acid saturation of the colloidal complex and the pH of the soil solution increases correspondingly.

Lime requirement (LR) of soils

The **lime requirement** of a soil can be defined as the amount of liming material that should be added to raise the soil pH to some prescribed value, usually 6.5.

The lime requirement of a soil can be determined by several methods such as Woodruff's buffer method, SMP buffer method, Mehlich – $BaCl_2$ – TEA buffer method. Most of the soil testing laboratories in India adopt Woodruff's buffer method.

The buffer is prepared by dissolving 10 g calcium acetate. 12g P-nitrophenol, 10g salicylic acid and 4g calcium hydroxide in 900 ml distilled water and then diluted to 1000 ml with distilled water. The buffer pH is adjusted to 7.0 by adding dilute sodium hydroxide solution.

The soil is equilibrated with the buffer solution. The drop in pH buffer after equilibration with soil form the basis of LR as the buffer gives a linear change of pH in relation to exchangeable H+ (meq. /100 g soil) in the pH range 6 to 7.

The LR value is calculate as: LR = (pH of buffer-pH of soil buffer suspension) x 10,000 kg$CaCO_3$ / ha.

The woodruff buffer appears to be too strong to indicate any change in pH of weekly buffered soils rich in aluminium.

The SMP method is commonly used for high lime requirement soils with pH < 5.8, high organic matter and high soluble aluminium. On the other hand the Mehlich buffer method is used for highly weathered and leached soils.

The amount of lime required to correct soil acidity depends on the texture and organic matter content of soil, composition and neutralizing value of the lime stone. The LR value for a coarse texture soil low in organic matter is less than a fine textured soil with high organic matter content to affect the same pH change. More lime is needed for strongly acid soils than for weakly acid soils. For a given pH, a sandy soil requires less lime than clay loam.

Types of liming materials

The materials commonly used for liming the soils are the oxides, hydroxides, carbonate and silicate of calcium or calcium and magnesium. The associated ions with Ca or Ca and Mg is very important since, it reduces the activity of the hydrogen ion. On this basis, gypsum ($CaSO_4.2H_2O$) does not alter the soil pH, although it is a source of Ca and S. However, sub soil acidity can be ameliorated with surface application of gypsum.

Various liming materials used in agriculture with their reaction in soil is described in the following paragraph.

1. Calcium oxide (CaO)

Calcium oxide is a white powder, known as un-slaked lime, burnt lime or quicklime. It is manufactured by heating calcite lime stone in an oven or furnace.

$CaCO_3$ + heat $\rightarrow$ CaO + CO_2

When added to soil, it reacts immediately as per the following reaction:

$$\text{soil}\left]\begin{array}{l}H\\H\\H\\Al\\Al\end{array}\right. + H_2O + 2CaO \rightarrow \left.\begin{array}{r}H\\Ca\\Ca\\Al\\H\end{array}\right[\text{soil} + Al(OH)_3$$

Complete mixing of CaO in soil is difficult since, it is a white powder. However, thorough mixing is necessary to avoid formation of granules or caking.

2. Calcium hydroxide

Calcium hydroxide ($Ca(OH)_2$) is a white, powdery substance frequently called as slake lime, hydrated lime or building lime. Addition of water to burnt lime makes hydrated lime, which is more caustic than burnt lime.

$$CaO + H_2O \rightarrow Ca(OH)_2$$

When added to acid soil it reacts as:

$$\text{Soil}\left]\begin{array}{l}H\\H\\H\\Al\end{array}\right. + 2Ca(OH)_2 \rightarrow \left.\begin{array}{r}H\\Ca\\Ca\\H\end{array}\right[\text{Soil} + Al(OH)_3 + H_2O$$

Calcium and Calcium - Magnesium Carbonate

Crystalline calcium carbonate is termed as calcite or calcite lime stone. Crystalline Calcium – Magnesium carbonate is known as dolomite. When applied to soil, these materials undergo following reaction:

$$\text{Soil}\left]\begin{array}{l}H\\H\\Al\end{array}\right. + 2CaCO_3 + H_2O \rightarrow \left.\begin{array}{r}H\\Ca\\Ca\end{array}\right[\text{Soil} + 2CO_2 + Al(OH)_3$$

$$\text{Soil}\left]\begin{array}{l}H\\H\\Al\end{array}\right. + CaMg(CO_3)_2 + H_2O \rightarrow \left.\begin{array}{r}H\\Ca\\Mg\end{array}\right[\text{Soil} + 2CO_2 + Al(OH)_3$$

Use of naturally occurring lime sources such as calcite, dolomite has limited use in agriculture because of high cost since these materials have industrial use.

Several industrial wastes such as paper mill sludge (PMS) from paper mills, basic slag from steel industry, press mud from sugar mills using carbonate process have been tested successfully as amendments for acid soils, which are ecofriendly. A suitable liming material must be locally available, properly ground and should have high neutralizing value and low cost for use by small and marginal farmers. Although huge amount of basic slag (100 milliontonnes per annum) is generated from steel plants located in Bhilai, Rourkela, Bokaro, Durgapur, Burnpur and Jamshedpur, its use is limited due to high grinding cost. On an average, Indian slag contains 1 to 3 percent P_2O_5, 24 to 50 percent CaO and 2 to 10 percent MgO. The different industrial byproducts used as liming materials in acid soils are discussed here.

i) Phosphogypsum

Phosphogypsum is a byproduct of diammonium phosphate fertilizer plant. It contains 16% S and 20-21% Ca. Since, Ca in phosphogysum leached down faster in light textured soils as compared to calcite, sub soil acidity could be ameliorated with resultant reduction of aluminium toxicity. It also helps in correcting S deficiency in acid soils. It is reported that application of phosphogypsum @ 60 Kg S ha^{-1} reduced the exchange acidity, exch. Al^{3+} and exch. H^+ from 0.467 to 0.344, 0.2 to 0.1 and 0.267 to 0.244 c mol (+) Kg^{-1}, respectively within 30 days of incubation period.

ii) Paper mill sludge (PMS)

Paper mill sludge is a byproduct of paper mill. The neutralizing value of PMS is 65-79% and Ca content 25 to 35%. Application of PMS to acid soils increased the soil pH and decreased the exch. Al^{3+} and H+. Application of PMS @ 0.2 LR is adequate to bring the soil pH to a safe level for profitable crop production with neutralization of total exch.Al^{3+} within first week of application.

iii) Press mud (PM)

Press mud is a byproduct of sugar industry and can be used as an amendment to improve soil physical properties. The pH of PM ranges from 6.5 to 6.92 and OC. 16-18%. The neutralizing value of 22-24% indicates that it cannot be compared with $CaCO_3$ or PMS as a liming material, but due to high organic carbon content, it can be used as an acid soil amendment.

iv) Stromatolyte

Deposits of 40 million tonnes of stromatolyte rock is reported in Koraput district of Odisha. The neutralizing value of stromatolyte ranges from 79-81% and calcium content 16 -17.7% and magnesium content 9.6 - 11.1 % (Table 6.8).

Table 6.8. Calcium, Mg and S content of stromatolyte.

Mesh size	N.V. (%)	Ca (%)	Mg (%)	S(%)
22	79.0	16.0	9.6	0.035
60	79.6	17.4	10.5	0.047
100	79.8	17.9	10.8	0.047
350	81.0	17.7	11.1	0.047

Several incubation and field studies indicated that application of stromatolyte (22 mash) @ 0. 5 LR to acid soil increased soil pH and decreased exch. H^+ and exch. Al^{3+} after 28 days of application.

v) LD slag

LD slag is a byproduct of steel industry (SAI, Rourkela) having pH 10.49, Ca content 16.82%, Mg content 2.13% and neutralizing value of 58.8 %. Several laboratory and field studies conducted in Orissa University of Agriculture and Technology, Odisha indicated that application of LD slag @ 0.3 LR increased the soil pH from 4.53 to 6.98. About 75% of exch. Al^{3+} was neutralized within 14 days of LD slag application. The combined application of LD slag with PMS in 1:1 ratio @ 0.3 LR was better than sole application of LD slag which increased the ground nut pod yield by 58% to 90%.

vi) Basic slag

Basic slag is a byproduct of steel industry and contains 30-45 % CaO and 2% total P. Use of basic slag in agriculture has been suggested as a P – carrier and as an amendment for acid soils. The beneficial role of basic slag in increasing crop production has been well established in acid soils of Bihar and Jharkhand on maize, wheat, gram and groundnut. However, there is limitation in use of basic slag because of high grinding cost and heavy metal content (Cd, Cr, Ni).

Neutralizing value of liming materials

The neutralizing value or calcium carbonate equivalent of liming materials differ markedly in their ability to neutralize soil acidity. The amount of liming material depends on its neutralizing value. The neutralizing value is related to the molecular composition of the liming material and its purity. The neutralizing value of pure calcium carbonate is taken as 100 against which other liming materials are measured.

The neutralizing value of liming materials is calculated by using the formula:

$$X = \frac{100}{\text{Mol weight of liming material}} \times 100$$

Where x, is the neutralizing value of a liming material

For an example : The neutralizing value of

$$CaO = \frac{100}{\text{Mol. wt. of CaO}} \times 100 = \frac{100}{56} \times 100 = 179\ \%$$

The neutralizing value of different liming materials are –

CaO	-	179 %
$Ca(OH)_2$	-	136 %
$Ca\ Mg\ (CO_3)_2$	-	109 %
$CaCO_3$	-	100 %

Fineness of lime stone

The efficiency of lime stone depends on its degree of fineness. The rate of reaction depends on the surface in contact with the soil. The rate of reaction is slow if the particles are coarse, but rapid in case of fine particles. However, the cost of lime stone increases with the fineness of grinding. Agricultural lime stones generally contain a mixture of both coarse and fine materials.

The efficiency of different sizes of lime stone are given below;

Size of materials	Efficiency rating
1. Materials passing through a 60 mesh sieve	100
2. Passing through 20 but not 60 mesh	60
3. Passing through 8 but not 20 mesh	20

For example- The efficiency rating of a lime stone containing 60% of 60 mesh size, 30% of 20-60 mesh size and 10% of 20-60 mesh size is calculated as;

60% through 60 mesh size = 60% x 100 = 60

30% through 20-60 mesh = 30% x 60 = 18

10% through 8-20 mesh = 10% x 20 = 2

Total efficiency rating of the lime stone is 80.

When 100 percent of a lime stone passes through a 60-mesh sieve, the efficiency rating is 100.

Liming programme

The primary objectives of liming are:

i) Increase pH of acid soils so that other plant nutrients will be available.

ii) Supply calcium and magnesium and

iii) Decrease the toxic effect of iron, aluminium and manganese.

A successful liming programme depends on:

i) Selection of liming material

ii) Dose, time, frequency and method of application

iii) Crop response to liming

The availability of cheap and effective liming material is very important for the success of acid soil amelioration programme. The commercial grade of liming materials such as calcite, quick lime, hydrated lime or dolomite are more reactive and efficient but, their use in agriculture is limited because of high cost.

A number of industrial by products such as basic slag, PMS, phosphogysum, press mud are being used for amelioration of acid soils in states of Odisha, Jharkhand, Chhattisgarh, West Bengal which are locally available at a low cost. Lime stone deposits in North-Eastern States, Assam, Himachal Pradesh are potential sources for producing agricultural grade lime at low cost. The corps like pigeon pea, soyabean, cotton responded well to liming. The crops grouped under medium responsive are chickpea, lentil, peas, ground nut, maize and sorghum and low responsive are small millet, rice, potato etc.

Several filed experiments conducted in acid soils of Jharkhand, Odisha, West Bengal indicate that application of lime @ 0.1 to 0.2 LR in lines below the root zone at sowing was found economical and beneficial for crops. The yield of groundnut, maize, soyabean etc. increased significantly with liming.

The efficiency of liming materials increases when it is applied with FYM or organic manure. Several studies indicated that lime with 50% recommended dose of fertilizer resulted in higher crop yield in acid soil regions.

Beneficial effects of liming

Liming benefits the soils in several ways:

1. Lime increases the availability of nitrogen by hastening the decomposition of organic matter. Nitrification and nitrogen fixation by legumes are encouraged by liming.

2. Lime makes phosphorus more available. In low pH soils, phosphorus is fixed by soluble Fe and Al. Liming reduces the solubility of Fe and Al by increasing soil pH, and therefore less phosphorus is held in insoluble and unavailable forms.

Application of excessive amounts of lime greatly increased the soil pH and thereby phosphate availability will again be decreased because of precipitation as calcium or magnesium phosphates. A liming programme should be planned to keep the soil pH between6 to7 for maximum benefit.

3. Lime makes potassium more efficient in plant nutrition. Lime reduces the excessive uptake of potassium which can lead to luxury consumption. In limed soil, higher uptake of calcium reduces excess uptake of potassium.
4. Lime furnishes calcium and magnesium (if lime is dolomite) for plant nutrition.
5. Toxic effect of iron, aluminium and manganese reduced with liming.
6. Molybdenum availability increases with increasing pH beyond 7.0. But, the availability of other plant nutrients affected at this pH range.
7. Over liming may leads to

i) Boron deficiency

ii) Fe, Mn, Cu or Zn deficiency

iii) P-availability may reduce to a critical level

iv) K uptake may be reduced.

Crop response to liming

Lime requirement of acid soils in India varied from 2.6 to 24.0 t $CaCO_3$ per hectare and was closely related to soil pH and organic matter content. Several studies indicated that full dose of lime application is often not economical.

Field experiments conducted in different districts of Odisha with varying pH levels using PMS @0.2LR, applied in furrows below the seeds at the time of sowing showed the beneficial effect of lime on several crops. The yield of groundnut increased by 17-36% over farmer's practice in red soils of Mayurbhanja, Ganjam and Nayagarh. The yield of green gram increased by 5-21% in laterite soils and cabbage and cauliflower by 15-22% in red soils (Table 6.9).

Table 6.9. Response of different crops to liming in acid soils of Odisha.

Crop	District	pH Range	No. of trials	Yield (t ha^{-1})		Yield response (%)	B:C Ratio
				FP (Farmer's practice)	FP+ Lime		
Groundnut (Kharif)	Dhenkanal	4.0-6.3	9	0.84	1.14	36	2.34
	Mayurbhanja	4.8.5.2	11	1.00	1.76	17	1.75
	Nayagarh	5.5-5.7	10	1.07	1.27	19	1.20
	Ganjam	5.6-6.1	7	1.91	2.39	25	4.32
Greengram	Khurda	5.5-6.5	12	0.81	0.89	5	1.50
	Dhenkanal	3.8-6.0	10	0.82	0.99	21	1.60
Cabbage	Kandhamal	5.9-6.6	5	11.17	13.00	16	5.40
	Koraput	5.8-7.3	7	22.40	25.69	15	-
Cauliflower	Kandhamal	6.0-6.1	5	9.95	12.10	22	14.05

Source: Jena (2011)

A number of field experiments were conducted in different states of India with locally available liming materials and various crops under the ICAR network project on "Soil characterization and resource management of acid soil regions for increasing productivity" during 2000 to 2005. Locally available liming materials @ 2-4 q/ha was applied in furrows in direct seeded crops at the time of sowing. Results revealed that the response of lime over farmer's practice was fairly good in ground nut + pigeon pea cropping sequence in Odisha (49.4%) and Jharkhand (30%), maize in Himachal Pradesh (16%) and Meghalaya (30%), mustard (18%) and summer green gram (14%) in Assam, wheat (16%) in Himachal Pradesh (18%) and ground nut (18 %) in Maharashtra.

References

Jena, D. (2011) Acid soil Management in India – Challenges and Opportunities." In Strategies for Arresting Land Degradation in South Asian Countries. Published by SAARC Agricultural Centre, BARC Complex.

Mandal, S.C. (1997) Introduction and Historical review. In Acid soils of India (Edited: I.C. Mohapatra et al.), ICAR, New Delhi, PP:3-4.

Murthy, R.S., Shankaranarayana, H.S. and Hirekerur, L.R. (1976) Distribution, Genesis and Classification of Acid soils of India. Bulletin of Indian Society of Soil Sciene. 11:1-17.

Singh, M.V. (2006) Micro – and Secondary Nutrients Problems and their Management in Acid soils of India. Symposium onNutrient Management in Acid Soils organized by ISSS at OUAT, Bhubaneswar, Nov. 10-13.2006.

Study Questions

1. The term agricultural lime refer to what materials?
2. Distinguish between active acidity and potential acidity. Which of these two forms is measured when a pH determination is made?
3. What is meant by the term buffer and buffer capacity?
4. How the buffer capacity of soil is related to its lime requirement value?
5. Define the term lime requirement.
6. How is the lime requirement of a soil determined?
7. Define soil pH. Calculate the H^+ ion and OH^- ion concentration of a soil having pH 8.
8. Why gypsum is not considered as a liming material?
9. Why calcite is not popular among farmers as a liming material?
10. Define neutralizing value of a liming material. What is the neutralizing value of CaO?
11. The neutralizing value of a liming material is 80%. How many tonnes of this lime stone would be equivalent to 4 tonnes of chemically pure calcium carbonate?
12. What are indirect benefits of adding lime to soil?
13. What would happen if over dose of lime is applied?
14. The soil analysis report of two field soils are given below;

Soil property	Field	
	A	B
Organic matter (%)	0.8	3.1
Clay (%)	10.0	35.5
Type of clay	1.1	2.1
pH	5.0	5.0

Does the lime requirement value of both field soils will be same or different,Give reasons?

15. What are different sources of hydrogen ions in soil?
16. Solution A has a pH of 4.0 and solution B has pH 7.0. The active acidity of solution A is how many times greater than that of solution B?
17. What happens when $CaCO_3$ is applied to soil?
18. The lime requirement value of an acid soil is 5 ton/ha. How much CaO is needed?

19. What is acid rain?
20. Write short notes on
    i) Suspension effect
    ii) Junction potential
    iii) Lime potential
    iv) Phosphate fixation
    v) Pressmud
    vi) Fineness of lime stone
21. Give a brief account of secondary and micronutrients status of acid soil regions of India.

7

Problems and Management of Saline and Alkali Soils

Saline and alkali soils occur commonly under arid climate due either to the presence of an excess of soluble sodium salts or to the predominance of sodium among the exchangeable bases.

Saline and alkali soils have been classified as follows

1. Saline soils

Saline soils were originally called "**White alkali**" soils. Russians term them as "**Solonchak**" soils.

The soils are classified as saline if the electrical conductivity of saturated soil paste is more than 4 dS /m at 25^{0}C, exchangeable sodium is less than 15 percent and pH is less than 8.5.

2. Alkali soils

The soils have been classified as alkali if the electrical conductivity of saturated soil paste is less than 4 dS/m, exchangeable sodium is more than 15 percent and pH is 8.5 to 10.0.

3. Saline-alkali soils

The term "Saline -alkali" applies to soils which are both saline and alkaline. Usually the soils have following characteristics:

i) The conductivity of saturated soil paste is greater than 4 dS/m at 25^{0}C.

ii) Exchangeable sodium is more than 15 percent.

iii) The soils have variable pH, usually above 8.5 depending on the relative amount of exchangeable sodium and soluble salts.

When exchangeable sodium will leach downward, the pH rises above 8.5, but the soluble salts accumulate again, the pH may again fall to 8.5.

Saline soils

Formation

In arid regions, due to low rainfall and high temperature, there is a tendency of accumulation of soluble salts near the soil surface. During rainy season, these salts move down ward to the lower layers but, during winter and summer season the soluble salts accumulates at surface due to intense evaporation. Repetition of this leaching-deposition processes results in a complex pattern of soils rich in soluble salts.

The salts mass have originated from chemical weathering of rocks and have been dissolved by percolating water. The usual source of origin is from saline ground waters either in the vicinity of inland seas or salt lake or in depressions where the water table is at near surface.

The favourable conditions for formation of saline soils are:

i) High water table with high salt content

ii) High temperature and

iii) Low rainfall.

Under irrigated conditions, saline soils have developed with high water table may be due to:

i) Excess application of irrigation water

ii) Irrigation with poor quality water

iii) Poor drainage keeps the salts in the surface soil and prevents the leaching of salts.

Coastal saline soils are recognized by high plant mortality and often by white salt crusts on the soil surface. Frequent inundation of land during high tide with sea water, ingress of sea water through drains, creeks and rivers cause salinization of land. Under ground water table is shallow and enriched with high salt content. During dry period, the salt moves along with water to the surface by capillary action. The water evaporates and leaves salts on the soil surface, which accumulates in toxic amounts that restricts plant growth during summer season.

Distribution of saline soils

Soil salinity is a major factor contributing the loss of productivity of cultivated lands. About 932.2 million (M) hectare of land in Globe are affected due to soil salinity and sodicity. These soils are seen in North America (15.8 M ha), Central America (2 M ha), South America (85.1 M ha), North and Central Asia (20 M ha), Europe (30.7 M ha), Australia (357.4 M ha).

It is estimated that about 20% (45 M ha) of irrigated land, producing one third of the world's food is salt affected. The amounts of World agricultural land degraded by salt accumulation each year is estimated to be 10 M ha. This rate can be accelerated by climate change, excessive use of ground water (mainly nearer to sea), increasing use of low quality water in irrigation and poor drainage. It is estimated that by 2050, 50% of the World's arable land will be affected by soil salinity.

In India about 7 M ha of land are salt affected out of which coastal saline soils occupy 2.1 M ha. The salt affected soils in India are seen in states of Uttar Pradesh (1.295 M ha), Gujarat (1.214 M ha), West Bengal (0.850 M ha), Rajasthan (0.728 M ha), Punjab (0.688 Mha), Maharashtra (0.534 M ha), Haryana (0.526 M ha), Odisha (0.404 M ha), Karnataka (0.404 M ha), Madhya Pradesh (0.224 M ha), Andhra Pradesh (0.042 M ha) and other states (0.040 M ha.).

Characteristics of saline soils

Saline soils are characterized by high concentration of soluble salts that significantly reduces the yield of most crops. The electrical conductivity of soil saturated paste (ECe) is more than 4 dS/m at 25^{0}C which are equivalent to approximately 4.0 mMNaCl and osmotic pressure of approximately 0.2 MPa. The pH of saline soils is generally below 8.5. Salts generally found in saline soils include chloride and sulphates of Na, K, Ca and Mg. Nitrate may be present in appreciable quantities only rarely. Concentration of boron above 1.5 ppm is unsafe for plant growth. Sodium and chloride are dominant ions and sodium contributes more than 50% of cations. Calcium and magnesium are usually present in sufficient quantities to meet the nutritional needs of crops.

Soil texture is an important criterion to characterize saline soils. Sandy soils with 0.1% salt concentration cause injury to the growth of common crops while the crops grow normally in clayey soils with the same salt content.

Excess salts in saline soils keep the clay in a flocculated state so that the soils generally have good physical properties. Soil structure is generally good and tillage characteristics and permeability to water are even better than those of non – saline soils.

The saline soils in Odisha occurs in a narrow strip of land adjacent to the coast of Bay of Bengal, which runs about 375 km long and ranges in width from 1-8 km. These soils are seen in low-lying land of river deltas, estuaries and depressions close to the coast. These soil are generally heavy textured with high clay content. Due to heavy texture, the soils have high water holding capacity and low infiltration rate and poor drainage. The ECe of surface soils varies from 6-10 dS/m and in some places reaches as high as 30 dS/m in month of May. The soluble salts comprises of chloride and sulphate of Na, Mg and Ca in decreasing order. Bicarbonate salts are present in traces and carbonates are absent. The predominance of sodium and magnesium both in solution and exchange phases along with high clay content makes the soil sticky in wet condition and become hard with wider cracks during dry period.

Besides salinity, lack of good quality irrigation water is another obstacle for successful rabi cultivation in coastal saline tract. The high ground water table along with low hydraulic conductivity and flat topography resulted in poor drainage and thus limits the scope of salt leaching. The typical characteristics of a saline soil in Odisha is presented in table 7.1.

Table 7.1. Physico–chemical properties of a saline soil profile (verticHalaquept), Motto, Balasore, Odisha.

Depth (cm)	Sand (%)	Silt (%)	Clay (%)	Textural Class	pH	EC (dS/m) (1:2.5 soil water)
0-20	13.5	40.0	46.5	C	7.1	13.5
20-65	12.0	36.5	51.5	C	7.8	8.4
65-95	17.0	46.5	36.5	CL	8.0	7.1

Source: Jena (1991)

Crop production constraints

The various problems associated with saline soils that interfere with plant growth are:

1. Soils are generally barren but, potentially productive. Non uniform crop stand, deep green, stunted plant growth.
2. Saline soils have a high wilting point and low amount of available moisture.
3. Soils have good permeability.
4. Sodium, calcium, magnesium are the main soluble cations and chloride and sulphate are dominant anions.
5. Excessive salts in the soil solution increase the osmotic pressure of soil solution compared to cell sap. It makes for plant roots to extract moisture due to higher potential force that holds water. If the salt concentration in soil is greater than that of the plant, water moves from the plant into the soil (**Plasmolysis**), which leads to wilting or death of the plant.
6. High concentration of soluble salts cause root injury and inhibition of seed germination.

Effects of salinity stress on plant growth

Saline soils containing high level of soluble salts affect plants by two primary mechanisms.-**osmotic effects and specific ion effects**. Soluble salts lower the osmotic potential of water making it more difficult for roots to remove water from soil. Plants spend more energy per unit quantity of water uptake. Higher concentration of certain ions including Na^+, Cl^-, HBO_4^-- and HCO_3^- are quite toxic to many plants. In addition to specific ion toxic effects, high level of Na^+ can cause imbalance nutrient uptake.

Several nutrients deficiency and nutritional imbalances may be caused by higher concentration of $Na^+$and Cl^- in soil solution derived from ion competition such as Na^+/Ca^{2+}, Na^+/K^+, Ca^{2+}/Mg^{2+} ,Cl^-/NO_3^- in plant tissues. Calcium deficiency symptoms are common when Na^+ / Ca^{2+} ratio is high in soil water.

Visual symptoms of salt injury in plant growth appears progressively. The first symptoms of salt stress are wilting, yellow leaves and stunted growth followed by chlorosis of green plants, leaf tip burning, necrosis of leaves and older leaves display scorching.

During the development of salt stress within a plant, all the major processes such as photosynthesis, protein synthesis, enzyme activity and energy and lipid metabolism are affected.

Plants are more susceptible to salt damage at early stage of growth. Salinity may delay or even prevents the germination of seeds. Young seedlings may be susceptible to saline condition than the older plants of the same species which could tolerate.

Soil salinity and nutrient availability

Crop growth may be adversely affected by salinity– induced nutritional disorders. These disorders may result from the effect of salinity on nutrient availability, competitive uptake and transport within the plant. High concentration of Na^+ and Cl^- in saline soils may depress nutrient ion activities and produce extreme ratios of Na^+/Ca^{2+}, Na^+/K^+, Ca^{2+}/Mg^{2+} and Cl^-/NO^{3-}. As a result, the plant becomes susceptible to nutritional disorders that may result in reduced yield and quality.

Nitrogen

Addition of N usually improves plant growth and yield whether the crop is salt – stressed or not. Under moderate saline conditions, addition of N improved growth and yield of apple, bean, tomato, cowpea and corn. Several laboratory and green house studies indicated that salinity can reduce N accumulation in plants since, increase in Cl^- uptake decreased the shoot – NO_3^- concentration. However, the higher concentration of NO^-_3 in soil solution, decreased the uptake of Cl^-. Ammonium – N applied to maize, melon and pea plants were more sensitive to salinity than NO_3-N fed plants when grown on solution culture. Application of NO^-_3 – N increased the accumulation of K^+ in plants under saline condition.

Phosphorus

Application of phosphorus to saline soils increased crop growth and yield but, did not necessarily increase crop salt tolerance. Addition of phosphorus had no effect on salt tolerance when the soil salinity increased from moderate to high levels. Accumulation of P in crop plants decreased by 20-50% when the soil salinity increased from low to high levels. The competitive action between Cl and $H_2PO_4^-$ may influence P accumulation and uptake by plants. Salinity stress may increase P requirements of some plants.

Potassium

Maintenance of adequate levels of K^+ is essential for plant survival in saline environment. Potassium is the most prominent inorganic plant solute that makes a major contribution to the low osmotic potential in the roots which is a prerequisite for turgor-pressure-driven solute transport in the xylem and the water balance of plant. Under saline-sodic conditions high level of external Na^+ interfere with K^+ acquisition by the roots. The selectivity of the root system for K^+ over Na^+ must be sufficient to meet the levels of K^+ required for metabolic processes for regulation of ion transport and for osmotic adjustment.

Several studies on agricultural crops showed that K^+ concentration in plant tissue decreases as the Na-salinity or Na^+/Ca^{2+} ratio in the root media is increased. In contrast, other research has demonstrated that K^+ may be preferentially acquired and transported against a strong Na^+ concentration gradient. As a resultK^+levels in leaves increased with increasing NaCl – salinity.

The presence of adequate amount of Ca^{2+} in soil solution increases the K^+/Na^+ ratio. Increase in Ca^{2+} leads to reduction of K^+ leakage from root cells and more favourable for root K^+ status. As a result the beneficial effects of supplemental Ca^{2+} on K^+ status of salt-stressed plants are pronounced in the root tissue rather than shoots of crop plants.

Calcium

The deficiency symptoms of calcium like black heart of celery, internal browning of cabbage and cauliflower, cracking and storage disorder of apple are more pronounced under saline condition than normal soil. As the salt concentration in the root zone increases, plant requirement for calcium also increases. At the same time uptake of calcium from the soil solution decreased because of ion interaction, precipitation and increase in ionic strength.

Maize is a suitable test crop to study the competitive interactions between Na^+ and Ca^{2+} under saline condition. In high Na^+/Ca^{2+} substrates, maize is sensitive to growth inhibition. Water transport properties of maize roots are adversely affected by NaCl induced morphological changes. Calcium addition tended to prevent these changes and mitigated reductions in root hydraulic conductivity. Addition of gypsum to saline soils increased the yield of potato, bean, pigeonpea by decreasing Na^+ and Cl^- uptake and accumulation.

Magnesium

Calcium is strongly competitive with Mg^{2+} and the binding sites on the root plasma membrane have less affinity for the highly hydrated Mg^{2+} than Ca^{2+}. Thus, high concentrations of substrate Ca^{2+} resulted in increased leaf-Ca with a marked reduction in leaf-Mg under saline environment.

It has been known for several decades that solution with a Mg^{2+}/Ca^{2+} ratio greater than one, reduces the growth of maize. Calcium-induced Mg^{2+} deficiency has been observed in several crops.

Sulphur

Effect of salinity on sulphur nutrition in plants has not been studied in detail so far. Rather information are available how the plant responds to sulphate salinity as compared to chloride-salinity. Many crops are very sensitive to high chloride levels. Some species are more tolerant to sulphate-salinity than chloride salinity. It is suggested that for vegetable crops the salt tolerance would be 2 dSm^{-1} greater in a sulphate system than chloride system. Sulphate-salinity also reduces the selenium and molybdenum uptake by several crops.

Micronutrients

The availability of several micronutrients depends on the pH of the soil solution as well as the nature of binding sites on organic and inorganic particle surfaces. In saline and sodic soils, the solubility of micronutrients such as Cu, Fe, Mn, Mo and Zn is low and plants grown in these soils often show deficiency of these elements.

However, the relationship between salinity and trace element nutrition is complex and salinity may increase, decrease or have no effect on the micronutrient concentration in plant shoots.

Salinity-induced Mn deficiencyhas been reported in barley, bean, corn, pea and tomato shoots and Mn addition increased the salt tolerance of barley.

Plants grown in salt affected soils shows zinc deficiency. Zinc applications have been found to improve growth in salt-stressed plants.

Information on effect of salinity on Fe, Cu and Mo concentration in plants are inconsistent. Salinity increased the Fe concentration in shoots of pea, tomato, soyabean but decreased its concentration in shoots of barley and corn.

The optimal concentration range of plant available B is very narrow for most crops. Under arid climate, B toxicity is more concern than B deficiency which is very common in humid regions. In coastal saline soils, several crops suffer due to B toxicity might be due to high concentration of B in sea water.

Evaluation of saline soils

The availability of water in saline soil reduced due to high osmotic pressure (OP) of the soil solution. The saline soils are evaluated on the basis of percentage or parts per million (ppm) of salt in them. However, interpretation of such data required salt composition. Chloride system of salts is considered more toxic than sulphate salts. Equal amounts of salts are more toxic in sandy soils than in loam or clay soils. These variations are eliminated by evaluating the osmotic pressure of the soil solution in saline soils. Direct determination of the osmotic pressure of soil solutions is tedious. The osmotic pressure of a solution can be estimated from conductivity values by the relationship:

OP = 0.36 x ECe(dS m^{-1})when electrical conductivity is expressed in deci Siemens per metre and OP in bar.

There is no critical point of salinity where plants fail to grow. As the salinity increases growth decreases until plants become chlorotic and die.

Electrical conductivity standards developed by the United States Salinity Laboratory for evaluating saline soils are listed in table 7.2.

Table 7.2. Soil salinity classes and crop growth.

Soil salinity class	Conductivity of saturated extract (dS m^{-1})	Plant growth conditions
Non saline	0-2	Salinity effects mostly negligible
Slightly saline	2-4	Yields of sensitive crops may be restricted
Moderately saline	4-8	Yields of many crops are restricted
Strongly saline	8-16	Only tolerant crops yield satisfactorily
Very strongly saline	>16	Only a few very tolerant crops yield satisfactory

Determination of content of chloride and sodium is required for fruit trees. Chloride in excess of 10 meq / L in soil solution is toxic to plants. Boron is highly toxic to plants. Plants may be injured if the concentration of B in saturation extract is as low as 1 ppm.

Management of saline soils

Sustainable agricultural management of saline soils is principally based on two main approaches:

i) Prevention (to elude further salinisation processes) and

ii) Remediation measures

In selection of most appropriate management strategies several critical variables should be considered

i) Soil geochemistry (type of salt and concentrations)

ii) Ground water hydrology (Water table fluctuation)

iii) Climatic condition (rainfall, evapo- transpiration) and

iv) Plant species selection.

Soil Reclamation

Reclamation of saline soils is too difficult which requires salt removal from the root zone. Perhaps this is the most effective and long-lasting way to minimize or even eliminate detrimental effects of soil salinity. However, the process being slow and expensive requires large quantity of quality irrigation water and effective soil drainage. Methods commonly adopted to accomplish include following:

Scraping

Removal of salts accumulated on the soil surface by mechanical means only had limited success although some farmers have adopted this processes. Although this method has temporary effect on crop growth, the ultimate disposal of salts still poses a major problem.

Flushing

Washing away the surface accumulated salts by flushing water over the surface is some-times used to reclaim saline soils. But this method does not have much practical significance.

Leaching

This is one of the most effective method for removing salts from the root rhizosphere zone. Leaching is associated with ponding fresh water on soil surface and allowing it to infiltrate. For effective leaching, the salty drainage water should be discharge through sub surface drains that carry leached salts out of the area under reclamation.

The extent of Leaching Requirement (LR) depends on

i) Soil properties such as soil texture, structure, pore geometry and clay mineralogy

ii) Initial soil salinity and composition of soluble salts

iii) Quality of water used for leaching

iv) Leaching methods

v) Water- table depth

vi) Efficiency of drainage system and

vii) Salt tolerance of crops to be grown.

The **leaching requirement** is defined as the fraction of irrigation water that must be leached through the root zone to control soil salinity at any specified level. The LR is calculated as:

$$\mathbf{LR} = \frac{\text{Ddw.100}}{\text{Diw}} = \frac{\text{ECiw.100}}{\text{ECdw}} \text{ where,}$$

LR – leaching requirement in percentage of water applied that must leach through soil

Diw – depth of water applied

Ddw – depth of water leaving the soil

ECiw – Conductivity of irrigation water

ECdw – conducting of the drainage water leaving the root zone

Leaching Fraction (LF) is defined as the proportion of the depth of irrigation water that must pass through the root zone to maintain desired salt and water balance and to prevent yield reduction beyond acceptable level.

Leaching should be done when the soil moisture content is low and groundwater table is deep. The actual choice will however depend on the availability of water and other considerations. In some parts of India, leaching is done during summer months because this is time when the soil is dry and water table is deepest.

In coastal areas leaching of salts in sub-soil possesses another constraints because of clay dispersion and sodicity. Therefore saline-sodic and /or sodic sub-soils require an application of appropriate amendments for Na replacement during leaching. Application of gypsum to pedosphere rich with excessive concentrations of salts has multiple beneficial effects. Addition of Ca helps to maintain microaggregate integrity in the soil profile. Application of gypsum improved physical properties like decreased clay dispersion, increased hydraulic conductivity and ensure better aeration.

Example: Calculate the leaching requirement to prevent the build-up of salts in the upper 30 cm if ECiw is 2.0 dSm^{-1} and ECdw is 5.0 $dSm^{-1.}$

Ans.

Leaching Requirement

$$LR = \frac{ECiw}{ECdw} X\ 100 = \frac{2}{5} \times 100 = 40\%$$

That means 40% of applied irrigation water with EC 2.0 dSm^{-1} should pass beyond 30 cm to control the buildup of salts beyond 5.0dSm^{1}.

Salt balance

Salt balance is used as an indicator of soil salinity. The salt balance indicates that input and output of all salts for a defined volume or depth of a soil profile. The general salt balance equation per unit area is given as:

[dw (VrCr + ViCi+VgCg) + Ms + Ma]–(Mp +Mc+dwVhCh) = DMsw Where,

VrCr	=	Volume of rainwater entering into soil with a salt conc. Cr
Vici	=	Volume of irrigation water into soil with a salt conc. Ci
VgCg	=	Volume of ground water with salt conc. Cg. entering the root zone by capillary rise
VhCh	=	Volume of water drained from the root zone with salt conc. Ch
Ms	=	Mass of salt dissolved from soil
Ma	=	Mass of salt added through input (fertilizer)
Mp	=	Mass of salt precipitated (adsorbed) in situ
Mc	=	Mass of salt removed by crop
dw	=	Density of water
DMsw	=	Change in mass of salts in soils liquid phase during the period under consideration

The salt balance equation can be simplified for a complete annual period. Assuming that the net change in total soil water content from one year to the next is closed to be zero. The water balance per unit area, disregarding surface run off, is as follows:

Vi + Vr + Vg = Vet + Vh Where,

Vi, Vr, Vg, Vet and Vh are total annual volume of irrigation, rainfall, capillary rise of ground water, evapotranspiration and drainage, respectively.

Since, ET removes no salt and crop generally removes only a negligible amount and if we disregard agricultural inputs and in situ precipitation and dissolution of salts, then the equation becomes:

ViCi + VrCr = (Vh–Vg) Chwhere, Ci, Cr, Ch are average concentration of salt in irrigation, rainfall and drainage water, respectively.

If the water table is kept very deep, no substantial capillary rise of groundwater into the root zone takes place and salt concentration of rainwater negligible, then the above equation becomes , ViCi = VhCh

Salt tolerance of crops

The salinity tolerance of any crop is defined as the ability to withstand or tolerate the effects of excess salt in the root zone. Salt tolerance is described by the decrease in relative yield with increase in soil salinity as –

$Y = 100 - (ECe - EC_t)$ Swhere,

Y = relative crop yield, 100 is the maximum yield

ECe-salinity of soil saturated paste

EC_t -the threshold value of electrical conductivity that is expected to cause significant reduction in the maximum expected yield

S – slope that represent the percentage of yield reduced for each unit of added salinity above EC_t.

The majority of vegetable crops have a salinity threshold of less than 2.5 dSm^{-1} (Table 7.3). The salt tolerance of crops depends on many factors such as plant growth stage, climate, salt type, soil properties.

In general, plants at seedling or establishment stage are more sensitive to salt stress than at later stages. During germination and emergence, the salt tolerance is based on percent survival, while during later stages, determination of tolerance is based on relative growth reduction. The classes of tolerance are sensitive, moderately sensitive, moderately tolerant, tolerant and unsuitable for crops (Table 7.3).

Table 7.3. Salt tolerance of vegetable crops as determined by soil salinity (ECe) and irrigation water salinity (ECw).

Crops	Soil threshold ECe (dS m^{-1})	Irrigation water threshold ECw(dS m^{-1})	Rating
Asparagus	4.1	2.7	T
Bean	1.0	0.7	S
Broccoli	2.8	1.9	MS
Carrot	1.0	0.7	S
Onion	1.2	0.8	S
Potato	1.7	1.1	MS
Spinach	2.0	1.3	S
Strawberry	1.0	0.7	S
Tomato	2.5	1.7	MS

S – Sensitive, MS – Moderately sensitive, MT – Moderately tolerant, T – Tolerant

The relative salt tolerance of different crops under Indian conditions are given in table 7.4.

Table 7.4. Relative tolerance of important plant species under Indian conditions.

Plant species	Threshold salinity (dS/m)	Plant species	Threshold salinity (dS/m)
Cluster bean	8.8	Maize	1.7
Barley	8.0	Sugarcane	1.7
Cotton	7.7	Tomato	2.5
Sugarbeet	7.0	Cucumber	2.5
Sorghum	6.8	Spinach	2.0
Wheat	6.0	Cabbage	1.8
Soyabean	5.0	Potato	1.7
Cowpea	4.9	Onion	1.2
Groundnut	3.2	Carrot	1.0
Rice	3.0	Turnip	0.9

Source: Singh (1998)

A classification of plants according to their salt tolerance has been made by the staff of the United States Salinity Laboratory at Riverside, California. The list of different crops namely, field crops, forage crops, fruits and vegetables are listed in table 7.5. Of the field crops, barely, sugar beet and cotton are the most tolerant to salt, while beans are most sensitive. Sweet clover and alfalfa (Lucerne) are intermediate to salt tolerance.

Date palm is the only fruit crop known to be very tolerant to salinity. Most fruits are sensitive to salt concentration in soil.

Garden beet, asparagus and spinach are high salt tolerant while radish, celery and garden beans are sensitive.

Table 7.5. Relative tolerance of crops to salinity.

Type	High salt tolerance	Medium salt tolerance	Low salt tolerance
1. Field crops	Barley, sugar beet, cotton	Wheat, rice, sorghum, maize, sunflower, castor	Bean (Field)
2. Forage crops	-	Sweet clover, alfalfa	White clover
3. Fruits	Date palm	Pomegranate, fig, olive, grape	Peas, apple, orange, straw berry, lemon, avocado
4. Vegetables	Beet, asparagus, spinach	Tomato, broccoli, cabbage, cauliflower, sweet corn, carrot, onion, peas, cucumber	Radish, celery, green beans

Rice in saline soils

Rice is not tolerant to excess salinity, but preferred in saline soils among other tolerant crops. The system of rice cultivation is advantageous to the crop rather than tolerance of the crop to soil salinity. Maintenance of submergence or standing water throughout the growing season brings about a significant reduction in root zone salinity either by leaching and / or dilution of salts. During rainy season rice is an important crop in coastal areas. Although the salinity is very high initially, after one or two rains salinity is reduced in root rhizosphere zone enabling planting of seedlings grown in relatively good soil. Rice gives satisfactory yield even when the ECe of soil is 15-20 dS/m. Rice can tolerate a high concentration of salts at germination but, sensitive at early stages of growth. The salt tolerance increases with age of the crop. Transplanting of older rice seedlings (30-40 days old) can over-come the effect of salinity during initial growth stage. The tolerance of rice decreases from panicle initiation stage to flowering stage. The salinity stress at this stage invariably results in reduced grain yield. Severe salinity delayed the flowering and the number of sterile spikelets increased.

In coastal regions of India, salinity is a perpetual problem due to inundation from the sea and intrusion of sea water through rivers or estuaries. Several tall indica and photo sensitivity rice varieties have been developed for coastal areas. Some of the saline tolerant rice varieties are MCM1, MCM2 (Andhra Pradesh), Pokhali (Kerala), SR26B and lunishree (Orissa), Getu, Damodar and Dasal (West Bengal). These varieties have a low yield potential compared to the dwarf high – yielding types.

In recent years, systematic breeding efforts have been made to release saline tolerance varieties. Some of the tolerant rice genotypes with high yielding characters were used extensively in hybridization programme as donors for salinity tolerance. Apart from hybridization, mutation breeding approaches were tried and some promising cultures, Mut- 1 (CSR4), evolved from high rice yielding variety IR 8.

Water management

Under saline conditions, the crop yield increased with proper water management. At saturation point the salt concentration or the osmotic pressure of soil is minimal. As the soil progressively dries, the salt concentration and osmotic pressure of soil increases making the soil water difficult to be absorbed by plants. For these reasons crops grown in saline soils must be irrigated more frequently compared to crops grown under non-saline conditions.

Irrigation method can play an important role in controlling salts in the root zone. Sprinkler irrigation is an ideal method for irrigating frequently and with small quantities of water at a time. Leaching of soluble salts is more efficient when the water application rate is lower than the infiltration capacity of soil and such condition cannot be achieved by flood irrigation method.

In drip irrigation method, water is supplied continuously at a point source and in the immediate vicinity of plant roots. This method is useful when irrigating with water of high salinity. This method has the advantage that it keeps the root zone moist therefore, maintaining a low salt level.

Nutrient management

Apart from the effect of water availability to plants the possible toxic effect of excess soluble salts in soils may interfere with the normal nutrition of crops in saline soils. At moderate saline conditions in soil solution, the plants generally try to exclude unwanted ions as far as possible and promote uptake of nutrients. With increasing the salt concentration the uptake of sodium and chloride ions increase. Excessive uptake of Na^+ and Cl^-, often results in reduced uptake of some essential plant nutrients causing nutrient imbalances and deficiencies. Thus, although the available status of nutrient in a soil might not be in a deficient range, its application might compensate for the reduced uptake by plants resulting from the antagonistic effect of excess uptake of certain ions.

Supplementary application of calcium may results in improved K and Ca uptake, dry matter yield and reduced accumulation of Na. Increasing the K and P in the rhizosphere solution reduced the accumulation of Na and Cl in plants.

Application of Zn fertilizer may be beneficial in saline environment. Application of $ZnSO_4$ improves salt tolerance of cereals and reduced uptake of Cd.

In arid regions, nitric acid and sulphuric acid fertigation reduces soil pH and increases Ca^{2+} dissolution in clay soils, thereby minimizing salinity injury due to

Ca^{2+}/Na^{+} competition. It may also reduce chloride salinity in the root zone, because the nitrate can counter balance the excess of chloride. Iron must be supplied in chelated form (Fe-DTPA, Fe-EDDHA) to increase its availability to plants.

The addition of NO_3^-, Ca^{2+}, K, P, salicylic acid and silicon to saline medium or in foliar application has improved salt tolerance of vegetable crops such as tomato, melon, bean, strawberry etc. Silica has been shown to be effective in mitigating salinity effects on several vegetable crops. Silica decreases the root-to-shoot translocation of Na^+, Cl^- and B in tomato grown on a sodic-boron toxic soil.

Biofertilizers

The use of biofertilizers can mitigate salinity effects on vegetables and reduce soil salinization. Plant growth promoting rhizobacteria (PGPRs), endo- and ecto-mycorrhizal fungi, and many other useful organisms led to improved nutrient uptake, plant growth and plant tolerance to salt stress. The inoculation of seeds of various crop plants such as tomato, bean, lettuce with PGPRs can result in increased root and shoot growth, dry weight, fruit and seed yield and enhanced tolerance of plants to salt stress. Biofertilizer can reduce soil salinization by reducing application of fertilizers, improving soil fertility by fixing atmospheric nitrogen, solubilizing insoluble soil phosphates, and producing plant growth substances.

Management of coastal saline soils

The specific problems of coastal saline soils are:

i) Tidal water inundation during high tides

ii) Very poor drainage

iii) Water logging in monsoon season because of heavy rainfall

iv) High water table with saline ground water

v) The area is mono cropped due to high soil salinity and non-availability of good quality irrigation water during winter and summer season.

Based on these specific problems following management practices are recommended for coastal saline areas:

1. Land must be protected from inundation of saline water through construction of earthen embankment with manually operated sluice gate. The sluice gate will operate to remove excess rainwater and also to prevent ingress of sea water into country side during high tide.

2. Land should be properly levelled to prevent accumulation of rain water in low lying areas. Field level bunds of 25cm high are to be constructed around the field to retain rain water. The ponded water will dilute the salt concentration and also meet the crop requirement during dry period.
3. Main and secondary drains are to be excavated to drain out excess water from field to sluice gate.
4. Twenty per cent of the land should be earmarked for dug out pond at the farm level to store rain water for subsequent utilization as irrigation for rabi crops. This will also facilitate easy drainage in the field.
5. Use suitable rice variety depending on land situation. Delay transplanting to 7-14 days and transplant 30-35 days old seedlings with 3-4 seedlings per hill. This will help to reduce mortality of seedlings since, during transplanting the salinity is not reduced sufficiently.
6. During winter – summer season, growing of salt tolerant crops like barley, linseed, chilli, water melon are recommended to reduce evaporation and thereby minimize salinization. During summer, the surface salinity increases if the land is kept fallow.
7. Application of different organic amendments increased the grain yield of rice. In situ green manuring with dhaincha gave highest grain yield followed by rice-straw incorporation, water hyacinth, FYM (Table 7.6). Application of chemical amendments like gypsum, paper mill sludge also have beneficial effect on rice.

Table 7.6. Effect of organic amendments on yield of rice- linseed cropping sequence in saline soil (Motto, Odisha).

Treatments imposed on kharif rice	Rice yield (q ha^{-1})		
	1987		1988
	Rice (cv. IR56)	Linseed (Residual)	Rice (cv. CR 1009)
Control	34.35	1.69	41.90
Rice straw @ 10 t ha^{-1}	41.79	2.87	51.11
Water hyacinth @ 10 t ha^{-1}	41.87	2.27	51.27
Azolla @ 0.5 t ha^{-1}(dual cropping))	40.12	2.07	45.00
FYM @ 5t ha^{-1}	42.07	2.78	52.22
Rice husk	40.55	2.67	46.27
Dhaincha (Insitu green manuring)	43.97	4.35	57.46
CD (0.05)	3.77	0.85	3.12

* ECe of surface soil was 12.5 dSm^{-1}

Source: Jena (1991)

Sodic Soils

Sodic soils include soils containing exchangeable sodium in a quantity sufficient to interfere with the growth of most crop plants and not containing appreciable quantities of soluble salts. The exchangeable sodium percentage is greater than 15, and the conductivity of the saturation extract is less than 4 dS m^{-1}. The pH of sodic soil ranges between 8.5 to10.

Formation

1. The CO^{-}_{3} and HCO^{-}_{3} containing ground water is one of the chief contributing factors in the formation of sodic soils in USSR.
2. Sodic soils are also formed by desalination in the absence of enough divalent cations in some parts of Nile Delta.
3. In indogangetic plains of North India, sodic soils with Na_2CO_3 at the surface soil occur on about 2.5 million hectares in low – lying areas. Irrigation or high salt content ground water seems to cause the formation of these soils.

These soils are developed due to alternate wet or dry conditions and topographic situations. During wet season, water accumulates in low-lying areas. Due to high evaporation in summer season, the solution is concentrated and SAR of water increases resulting high level of exchangeable sodium and high pH. The displaced calcium precipitated at high pH. Repetition of this process for long periods resulted in formation of sodic soils.

4. In some areas, ground water contains high amount of sodium carbonate. Due to excess evaporation and precipitation, highly alkali soils develop if the inflowing water has a positive residual alkalinity.
5. Due to high pH and the dominance of sodium ions in sodic soils, part of organic matter dissolved in the soil solution and deposited as a thin film on the soil surface or on the surface of soil aggregates. These soils are designated as **black alkali** soils.
6. When the soluble salts are slowly leached from a saline – sodic soils, the colloidal particles dispersed, moved down ward and deposited in sub soil forming a dense sub soil horizon. On continued alternate- wetting and drying, the dense sub soils material develop columnar structures of 1 ½" to 2 " diameter and 4 to 6 inches in length. Soils having such subsoil structure are designed as **Solonetz**.
7. With extensive leaching of a saline – sodic soil in absence of Ca and Mg, part of Na is gradually replaced by hydrogen. The resulting soil may be slightly acidic, but contain enough Na to give an unstable structure. Such type of soil is designated as **degraded –alkali** soil.

Characteristics

Excess exchangeable Na has an adverse effect on the physical and nutritional properties of the soil with reduction in crop growth. These soils lack appreciable quantities of neutral soluble salts but, contain measureable to appreciable quantities of salts capable of alkaline hydrolysis i.e. sodium carbonate. Therefore, the electrical conductivity of saturation soil are often less than 4 dS m^{-1} at 25^0C.

The principal cause of alkaline reaction of soils is the hydrolysis of either the exchangeable cations or salts such as $CaCO_3$, $MgCO_3$, Na_2CO_3 etc. Hydrolysis of exchangeable cations takes place according to following reactions:

$$\text{Clay micelle}\begin{matrix}Na^+\\Na^+\end{matrix} + H_2O \rightleftharpoons \text{Clay micelle}\begin{matrix}Na^+\\H^+\end{matrix} + Na^+ + OH^-$$

In this reaction H^+ is inactivated by exchange adsorption in place of Na^+. The displaced Na^+ does not combine with, or inactivate OH^- ions which results in an increase in the OH^- ion concentration and increased soil pH.

Hydrolysis of compounds like $CaCO_3$ or $MgCO_3$ takes place according to the reaction:

$$CaCO_3 + 2H_2O \leftrightarrows Ca^{2+} + 2OH^- + H_2CO_3$$

In this reaction H^+ from water is inactivated through combination with carbonate to form weakly ionized carbonic acid. Hydroxyl ions are not activated through combination with Ca^{2+} resulting in an alkaline solution. The hydrolysis of $CaCO_3$ or $MgCO_3$, is limited due to low solubility and therefore they tend to produce a pH between 8.0 to 8.2. Soils containing measureable quantities of Na_2CO_3 may have a pH more than 8.2. The pH increases with increasing amounts of Na_2CO_3 and may be as high as 10.0 to 10.5. This is due to higher solubility of Na_2CO_3 and therefore the greater potential for hydrolysis.

Exchangeable Na^+ exerts several effects on productivity of soils. Sodium ions when coupled with high pH have toxic action on roots.The soils tend to become more dispersed resulting breakdown of soil aggregates and lowers the permeability of the soil to air and water. Dispersion also results in the formation of dense, impermeable surface crusts that, hinder the emergence of seedlings.

High levels of exchangeable Na^+ promote deflocculation and swelling of clay particles. Coarse textured soils are generally less affected than are fine textured soils. Soils containing expanding types of clay, such as montmorillonite, swell and disperse more with sodium adsorption than do soils containing with nonexpanding clays. The degree of swelling is related to the specific surface area of soils.

Increase in ESP beyond 15% results decrease in infiltration rate, soil water storage following infiltration and hydraulic conductivity. Under field condition, water transpiration rate decreased due to blocking of soil pores by dispersion of clay particles at higher Na content.

Sodium, molybdenum and boron will be toxic in many places. In some places Na induced Ca deficiency may affect crop growth.

The sodium adsorbed by clay and organic colloids causes dispersion of clay results in a loss of desirable structure and the development of a puddle effect. High amount of sodium compared to calcium increases the zeta potential of the soil exchange complex which causes repulsion of clay particles from each other causing dispersion.

Such effects on physical properties reduce drainage and microbial activity. The high pH in sodic soils causes a reduction in the solubility and availability of iron, copper, manganese and zinc to plants.

Evaluation of sodic soils

1. Measurement of soil pH

The distinguishing characteristics of a sodic soil are high content of exchangeable sodium and low permeability.

Determination of exchangeable sodium is too time consuming to be used for routine evaluation of large number of samples. The pH test is a significant diagnosis of salt affected soils, but the relationship between pH and exchangeable sodium is complex. The pH of the saturation paste is lower than the pH of 1:2 or 1:5 soil water suspension. The most common procedure followed is to determine soil pH first on a paste and then on 1:2 or 1:5 soil water suspension. The increase in pH between the paste and the suspension is used as a criteria of exchangeable sodium. The difference of about 1 pH unit between the two readings indicates that the soil contains more than 15 percent exchangeable sodium. This procedure has given useful results in some areas. It is suggested that a preliminary investigation be made for representative soils in each geographical area. If a satisfactory relationship is obtained, the more rapid pH test is used.

If the relationship is not satisfactory, the determination of cation exchange capacity and exchangeable sodium may be necessary to evaluate sodic soils.

2. Evaluating ESP

The cation exchange capacity (CEC) of soil refers to the capacity to adsorb or exchange the positively charged cations, such as calcium, magnesium, potassium, sodium etc. The various adsorbed cations can be exchanged one for another depending upon their relative concentrations in the soil solution or on the exchange complex and the valency and size of the cation involved.

Exchangeable sodium percentage (ESP) is, the amount of adsorbed sodium on the soil exchange complex expressed in percent of the CEC in milliequivalents per 100 g of soil.

$$\text{Thus, ESP} = \frac{\text{Exchangeable Na (me.100 g soil}^{-1})}{\text{CEC, me, 100 g soil}^{-1}} \times 100$$

The adverse effect of exchangeable sodium on the growth and yield of crops occurs according to the relative crop tolerance to excess sodicity. Based on the ESP values, various sodicity hazard classes are:

ESP	Sodicity hazard
< 15	None to slight
15-30	Light to moderate
30-50	Moderate to high
50-70	high to very high
>70	Extremely high

3. Sodium adsorption ratio (SAR) as an Index of sodicity

Experimental determination of ESP is tedious, time consuming and subject to errors. Incomplete removal of salt solution during the washing step of CEC determination can lead to high CEC and therefore low ESP estimates.

To overcome this difficulty, several workers prefer to obtain an estimate of SAR from an analysis of the saturated soil extract.

Sodium adsorption ratio, SAR, is defined by the equation:

$$\text{SAR} = Na^{+} \Big/ \sqrt{\frac{(Ca^{++} + Mg^{++})}{2}} \quad \text{Where,}$$

All concentrations are in mmol per litre. The SAR of a saturation extract or irrigation water is useful in evaluating ESP or estimating the sodium hazard when a specific irrigation water is used.

Reclamation and management

Amendments

Reclamation of sodic soils requires the removal of part or most of the exchangeable sodium and its replacement by the more favourable calcium ions in the root zone.

Chemical amendments for sodic soil reclamation can be broadly grouped into three categories as:

i) Soluble sources of calcium: Gypsum ($CaSO_4$. $2H_2O$), calcium chloride ($CaCl_2$) and phospho-gypsum (an industrial byproduct)

ii) Sparingly soluble calcium salts : Calcite ($CaCO_3$)

iii) Acids or acid forming substances: Sulphuric acid, iron sulphate, aluminium sulphate, lime – sulphur, sulphur, pyrite etc.

Use of lime stone in degraded sodic soils

The suitability of the amendments for reclamation of sodic soils will largely depend on the nature of the soil and cost consideration. Ground lime stone, $CaCO_3$ is effective only when the pH of soil is less than 7.0 because its solubility decreases rapidly as the pH increases (Table 7.7).

Table 7.7. Effect of pH on the solubility of $CaCO_3$ in water.

pH	Solubility of $CaCO_3$ mel^{-1}
6.21	19.3
6.50	14.4
7.12	7.1
7.85	2.7
8.60	1.1
9.20	0.8
10.12	0.4

Ground lime stone is suitable for reclamation of degraded sodic soils because that contain excess exchangeable sodium and also appreciable quantities of exchangeable hydrogen.

Lime reacts in soils according to the reaction:

Na, H-Clay micelle + $CaCO_3 \rightarrow$ Ca-Clay micelle + Na HCO_3

Gypsum

Gypsum ($CaSO_4$. $2H_2O$) is a white mineral that occurs extensively in natural deposits. Gypsum is soluble in water and reacts with both the Na_2CO_3 and adsorbed sodium as follows:

Na_2CO_3 + $CaSO_4 \rightarrow$ $CaCO_3$ + Na_2SO_4 (Leachable)

$$\begin{matrix} Na^+ \\ Na^+ \end{matrix} \text{Clay – micelle} + CaSO_4 \rightarrow Ca^{2+} \text{ clay micelle} + Na_2SO_4 \text{(Leachable)}$$

Calcium chloride

Calcium chloride ($CaCl_2.2H_2O$) is a highly soluble salt which supplies soluble calcium directly. Its reactions in sodic soil are similar to that of gypsum.

$Na_2CO_3 + CaCl_2 \rightarrow CaCO_3 + 2\ NaCl$ (leachable)

$\begin{matrix} Na^+ \\ Na^+ \end{matrix}$ clay micelle + $CaCl_2 \leftrightarrows Ca^{2+}$ Claymicelle + 2 NaCl (leachable)

Sulphuric acid

Sulphuric acid (H_2SO_4) is a corrosive liquid and is usually 95 per cent pure. Upon application to soils containing calcium carbonate, it immediately reacts to form calcium sulphate and thus provide soluble calcium indirectly.

$Na_2CO_3 + H_2SO_4 \rightarrow CO_2 + H_2O + Na_2SO_4$ (Leachable)

$CaCO_3 + H_2SO_4 \rightarrow CaSO_4 + H_2O + CO_2$

$\begin{matrix} Na^+ \\ Na^+ \end{matrix}$ - Clay micelle + $CaSO_4 \leftrightarrows Ca^{2+}$ Clay micelle + Na_2SO_4 (Leachable)

Iron sulphate and aluminium sulphate

Chemically these compounds are $FeSO_4.7H_2O$ and $Al_2(SO_4)_3$ .18 H_2O.

Both these solid granular materials have high degree of purity and soluble in water. When applied to soils, these compounds hydrolyses to form sulphuric acid and in turn supplies soluble calcium through its reaction with lime present in soil.

Chemical reactions involved are:

$FeSO_4 + 2\ H_2O \rightarrow H_2SO_4 + Fe\ (OH)_2$

$H_2SO_4 + CaCO_3 \rightarrow CaSO_4 + H_2O + CO_2$

$\begin{matrix} Na^+ \\ Na^+ \end{matrix}$ Clay micclle + $CaSO_4 \leftrightarrows Ca^{2+}$ clay micelle + Na_2SO_4 (Leachable)

Similar reactions will occur when aluminium sulphate is used as an amendment.

Sulphur

Sulphur is not soluble in water and does not supply calcium directly for replacement of adsorbed sodium.

When applied to sodic soils, the reactions occur as follows:

$2S + 3O_2 = 2\ SO_3$ (Microbial oxidation)

$SO_3 + H_2O \rightarrow H_2SO_4$

$H_2SO_4 + 2\ CaCO_3 \rightarrow CaSO_4 + Ca\ (HCO_3)_2$

$\begin{matrix} Na^+ \\ Na^+ \end{matrix}$ Clay micclle + $CaSO_4 \leftrightarrows Ca^{2+}$ clay micelle + Na_2SO_4 (Leachable)

The microbial oxidation of S is carried out by aerobic *Thiobacilli*group of Chemoautotrops.

Pyrite

Oxidation of pyrite (FeS_2) like sulphur leads to formation of sulphuric acid. The overall reactions are:

$$2\ FeS_2 + 7O_2 + 2H_2O \rightarrow 2\ FeSO_4 + 2\ H_2SO_4$$

The first step of reaction is non biological and iron (ferrous) sulphate is formed.

$$2\ CaCO_3 + H_2SO_4 \rightarrow CaSO_4 + Ca(HCO_3)_2$$

$$\begin{matrix} Na^+ \\ Na^+ \end{matrix} \text{Clay micclle} + CaSO_4 \leftrightarrows Ca^{2+} \text{ clay micelle} + Na_2SO_4 \text{ (Leachable)}$$

The rate of oxidation of pyrite is slow and depends upon the moisture content, oxygen availability, temperature and particle size.

Choice of amendment

The choice of an amendment will depend upon its relative effectiveness, improvement of soil properties and crop growth and the relative cost involved.

Instead of complete removal of sodium, it is often desirable to reduce the sodium to only 10 percent of the exchange capacity. This permits satisfactory crop growth, and natural processes associated with root activity, irrigation and leaching can complete the removal.

The objective of application of an amendment is to furnish soluble calcium to replace exchangeable sodium. For saline-sodic and nonsaline-sodic soils, lime stone is too insoluble to promote rapid reclamation. If sodic soils contain lime, then acidulating materials, such as sulphuric acid release calcium satisfactory.

Calcium chloride is the most readily available source of soluble calcium but, rarely been used for reclamation on an extensive scale because of its high cost. Similarly iron and aluminium sulphates are usually too costly and have not been used in large scale. The amendments like sulphur and pyrite must first be oxidized to sulphuric acid by soil microorganisms before they are available for reactions, the amendments are relatively slow actions. Being cheapest and most abundantly available, **gypsum** is the most widely used and usually gives best results.

Quantity of amendment

The quantity of an amendment necessary to reclaim sodic soil depends on the total quantity of exchangeable sodium to be replaced, relative tolerance of a crop and the soil depth upto which excess adsorbed sodium must be replaced.

If a quantitative exchange of applied soluble calcium for adsorbed sodium is assumed, replacement of each mole of adsorbed sodium per 100 g soil will require half a mole of soluble calcium. The quantity of pure gypsum required to supply half a c mole per kg of soil for upper 15 cm soil depth will be:

$$GR = \frac{\text{Molecular weight of gypsum}}{200} = \frac{172}{200} = 0.86\,g/kg\,soil$$

$$= 86 \times 10^{-5} \times 2.24 \times 10^{6}\ kg/ha$$

$$= 1926\ kg\ \text{or}\ 1.93\ t/ha.$$

The gypsum requirement (GR) for reclaiming sodic soil is determined in the laboratory. The test is performed by mixing a small soil sample (5g) with a large volume of saturated gypsum solution and measuring the calcium lost from the solution after reaction with soil. The decrease in calcium from the solution is expressed on the basis of tonnes of $CaSO_4 \cdot 2H_2O$ is called gypsum requirement of the soil.

The quantities of different amendments for reclamation of sodic soils relative to 1 ton gypsum are given in table 7.8.

Table 7.8. Equivalent quantities of common amendments for sodic soil reclamation.

Amendments	Relative quantity
Gypsum ($CaSO_4 \cdot 2H_2O$)	1.0
Calcium chloride ($CaCl_2 \cdot 2H_2O$)	0.85
Sulphuric acid (H_2SO_4)	0.57
Iron sulphate ($Fe\ SO_4 \cdot 7H_2O$)	1.62
Aluminiumsulphate $Al_2(SO_4)_3 \cdot 18\ H_2O$	1.29
ElmentalSulphur (S)	0.19
Pyrite (FeS_2) – 30% S	0.63
Calcium poly sulphide (CaS_5) – 24% S	0.77

Gypsum should be broadcast and thoroughly mixed with 0-10 cm soil by using a cultivator. Mixing of gypsum in deeper layer decreases its effectiveness, as a fraction of applied gypsum can react with soluble carbonates to precipitate Ca as $CaCO_3$. The gypsum ground to pass through 2 mm sieve has been found cost effective and also efficient.

Prevention of freezing is an important part of reclamation

The clay of saline-sodic soils is usually flocculated and influences the degree of swelling as long as the concentration of salts remain high. If such soils are leached with water of low salt content, dispersion and swelling of clay may seal off further movement of water. Such a transition is called **freezing**. Prevention of "Freezing" is an important part of reclamation.

Organic manures

Application of organic manures, rice straw incorporation, green manuring have beneficial effects on sodic soils. The decomposition of materials increases the concentration of CO_2 and organic acids in soil which help in inmobilizing calcium by dissolving calcium compounds.

Crops in sodic soils

Selection of crops during reclamation of sodic soils is important. Rice, grasses and dhaincha appear to be tolerant to sodic soils. Wheat, barley, oat, cotton, berseem, sugar cane and bajra are only moderately tolerant and legume crops are relatively sensitive to excess exchangeable sodium. It has been observed that, generally the crops that are able to withstand excess moisture conditions resulting in short term oxygen deficiency are more tolerant to sodic conditions.

Rice as a reclamative crop

Cultivation of rice under submerged condition (a layer of water on the field throughout the growing season) reduces soil pH and ESP due to continuous flooding. The low permeability of sodic soils is a further advantage to rice because of loss of water due to deep percolation are restricted. Apart from being tolerant to high sodicity, growing rice results in soil improvement through reduction in soil sodicity. Results of several studies showed that the growth of rice resulted in high removal of exchangeable sodium caused by mobilization of native insoluble $CaCO_3$ as a result of hydrolysis.

In conclusion its shallow and superficial root system, high sodicity tolerance and reclamative action together with storing of a large fraction of rain water makes rice an ideal crop during reclamation of sodic soils.

Grasses

Grasses are more tolerant to sodic conditions than field crops. Karnal grass (*Diplachne fusca*), Rhodes grass (*Chloris gayana*), Para grass (*Brachiaria mutica*) and Bermuda grass (*Cynodon dactylon*) are highly tolerant and can be successfully grown in sodic soils. In absence of any chemical amendments, karnal grass grows well in soils with high ESP (80-90). Karnal grass and Para grass are also highly tolerant to ponded water conditions and grow well in low lying sodic soils during rainy season. When grasses are grown, there is a continuous decrease in soil sodicity and an improvement in soil physical properties due to biological action of grass roots. In conclusion, growing tolerant grasses not only provide forage but also improve soils by increasing water absorption, reduced run off and soil losses due to erosion.

Trees

Studies showed that, there is a greater possibility for utilizing sodic waste lands for forestry. Forest species like *Eucalyptus* hybrid, *Prosopis juliflora* and *Acacia nilotica* could be grown successfully in highly sodic soils. The seedlings are to be planted in pits of 90 cm deep and 90 cm diameter after the pit soils amended with gypsum and manure. The survival of seedling can be increased to 100 percent if the tree seedlings are planted in auger holes of 15 cm diameter and 150 cm deep, filled with a mixture of original soil, 2 kg gypsum and 7-8 kg manure. In this technique a favourable environment is created for root growth by penetrating the sodicity and hard surface layers.

Crop growth stages and sodicity

Crops tolerance to sodicity changes with growth stages. The sodicity tolerance of rice increased with age and it is beneficial to transplant aged rice seedling of 35 to 40 days old in sodic soil. Many legumes germinate well in sodic soils but their subsequent growth is arrested due to low tolerance. Cotton is considered tolerant to salinity, but moderately tolerant to sodicity and sensitive at germination stage.

Fertilizer management in sodic soils

Nitrogen

Due to low organic matter content, sodic soils are generally deficient in available nitrogen. Presence of excess exchangeable sodium in exchange complex imparts structural instability to soil leading to poor physical properties. The infiltration rate of the soils is low with restricted internal drainage. The surface layers remained nearly saturated for prolonged periods following irrigation or rain resulting in temporary anaerobic conditions. Considerable losses of N in ammonia form due to volatilization are likely to occur in sodic soils due to high pH. High soil pH and poor soil physical conditions are likely to affect the transformations and availability of applied N fertilizers.

The complete hydrolysis of urea increased from about 3 days in a soil pH of 8.6 to about 7 days for soil pH of 9.8 leading to N losses. The activity of the urease enzyme decreased with high pH leading to delay in urea hydrolysis. The results of field studies showed that rice and wheat responded well to higher dose of N in sodic soils. Based on the results, it is generally recommended that crops grown in sodic soils be fertilized at 25 percent excess over the rates recommended for normal soils. Additional dose of N increased uptake of calcium and magnesium and decreased uptake of sodium.

Phosphorus

Sodic soils are generally rich in soluble P. Presence of sodium carbonate in these soils resulted in the formation of soluble sodium phosphate and hence have a positive correlation between electrical conductivity and soluble P status. Several studies indicated that crops in freshly reclaimed sodic soils did not respond to P fertilizers for 4-5 years because of their high P status.

Potassium

Several studies have shown that increasing soil sodicity resulted in reduced uptake of potassium by most crops. The significance of reduced uptake of potassium with increasing ESP needs potassium fertilization for crops. In some sodic soils lack of response to applied potassium was noticed since, presence of K-bearing minerals in sodic soils could supply sufficient K to meet the crop requirement.

Calcium

Increased uptake of sodium and decreased uptake of calcium by plants is a general phenomenon in sodic soils. With increase in ESP, the increase in sodium concentration of plants is much larger compared to the decrease in the calcium concentration. For this reason, the plants often accumulate sodium in toxic amount before the calcium becomes limiting for plant growth. Application of gypsum is absolutely necessary when the soils contain appreciable quantities of free sodium carbonate and the soil pH is high.

Micronutrients

The availability of micronutrients in sodic soils is affected due to high pH, low organic matter content and presence of calcium carbonate. Zinc deficiency is very common in sodic soils and application of 10 kg $Zn\ SO_4$/ha was sufficient to mitigate the deficiency of Zn in rice grown in an amended, highly sodic soils.

Next to Zinc, iron is often limiting in sodic soils due to high pH and presence of calcium carbonate. Addition of iron salts to correct the deficiency was not effective. Research results showed that there was marked increase in Fe and Mn status of a sodic soil on submergence up to 60 days. The increase was more when organic materials like rice husk or farmyard manure were incorporated in the soil.

Boron and molybedenum are not likely to be limiting for plant nutrition in sodic soils. However, there was marked reduction in the water soluble boron content of a highly sodic soil upon addition of gypsum. At high pH and sodicity, bron is present as highly soluble sodium metaborate which upon addition of gypsum is converted to relatively insoluble calcium metaborate. As with B, solubility of Mo increases with pH and forage crops grown on sodic soils accumulate Mo in excessive quantities, which may be toxic to the animals feeding on them.

References

Jena, D. (1991) Soil related constraints in coastal saline soils of Orissa. J. Indian Soc. Coastal Agric. Res. 9(1/2); 183-186

Singh, N.T. (1998) Water quality guide lines and tolerance limits for crops. In Agricultural salinity Management in India. CSSRI, Karnal.

Study Questions

1. Write short notes on

i) Osmotic potential

ii) Leaching Requirement (LR)

iii) Salt tolerance of crops

iv) Black alkali

v) Solonetz

vi) Degraded alkali

vii) Exchangeable sodium percentage (ESP)

viii) Sodium absorption ratio(SAR)

ix) Residual sodium carbonate (RSC)

x) Gypsum Requirement (GR)

2. Why do sodic soils occur in dry regions?
3. Differentiate between inland saline and coastal –saline soils.
4. Why do plants grow in saline soils appear water stressed?
5. Differentiate between saline and alkali soils.
6. What causes dispersion of sodic soils?
7. Why gypsum is most commonly recommended for reclamation of sodic soils? Give its chemical reaction in sodic soils.
8. Calculate the leaching requirement (LR) to prevent buildup of salts in the upper 30 cm soil depth if the ECiw is 1.5 dSm^{-1} and ECdw is 6.0 dSm^{-1}.
9. Why reclamation of coastal saline soil is difficult?
10. Briefly discuss the role of potassium in crop nutrition in saline soils.
11. Discuss about the calcium and magnesium interaction in saline soils.

12. Why rice crop grows well in saline environment although, it is not tolerant to excess salinity?
13. What types of management practices you would recommend for coastal saline soils?

8

Quality of Irrigation Water

Water used for irrigation contain some salts, but quantities and kinds vary greatly depending on the source of water. Since sodic status of soils can be greatly influenced by salts in irrigation waters, successful irrigation practices require the evaluation system to judge irrigation water quality and methods employed. The actual damage done by salt to plants or soils depends on the concentration in soil solution rather than the quality of irrigation water applied. Thus, the use of some water source for irrigation might lead to severe salt problem in soils where drainage is restricted.

Criteria for evaluation of irrigation water

Various criteria considered in evaluating the quality of irrigation water are:

1. Salinity hazard

The most satisfactory method for rating the salt content of irrigation water involves measuring electrical conductivity. Various conductivity limits have been proposed to indicate the degree of salinity problems that might be anticipated with use of irrigation water on crops. There are four classes of salinity hazard such as C_1, C_2, C_3, and C_4.

C_1 - **Low salinity water**: This water can be used for irrigation on all soils and on most crops but leaching is required in case of extremely low permeability soils.

C_2 - **Medium salinity Water**: This water can be safely used for moderate salt tolerance crops. The soils should have moderate level of permeability and leaching to avoid accumulation of salts.

C_3 - **High salinity water**: This water can not be used on soils with poor drainage. For use this water, drainage should be improved and salt tolerance crops should be selected.

C_4 - **Very high salinity water**: This water is unfit for irrigation on ordinary conditions. This water can be used occasionally if soil is permeable with good drainage facility.

2.Sodium hazard

Irrigation water containing high sodium is undesirable because sodium adsorbed on the soil cation exchange sites will leads to break down of soil aggregates, clogging the pores and making it impermeable to water flow. Based on the SAR of irrigation water, it is grouped into four classes as S_1, S_2, S_3 and S_4.

S_1 - **Low sodium water**: This water can be used for irrigation on all most all crops with very little danger of accumulation of sodium at the harmful level.

S_2 - **Medium sodium water**: This water can be used only on coarse textures soils with good permeability. It poses harmful effects if applied to fine textured soils with low permeability.

S_3 - **High sodium water**: This water have harmful effects on level of exchangeable sodium in most soils. It can be used under good drainage condition along with addition of organic matter and gypsum.

S_4 - **Very high sodium water**: This water is not fit for irrigation except at low and medium salinity level. Gypsum should be applied to decrease sodium hazard appreciably in case this water is used for irrigation.

Based on the salinity hazard and sodium hazard the irrigation water is classified as:

Water Quality hazard	Salinity hazard (EC dS m^{-1})	Sodium hazard (SAR)
Low	0.1 – 0.25 (C_1)	0 -10 (S_1)
Medium	0.25-0.75(C_2)	10-18 (S_2)
High	0.75-2.25 (C_3)	18-26 (S_3)
Very High	>2.25 (C_4)	> 26 (S_4)

3. Salt index

Salt index is used to evaluate the quality of irrigation water. It is calculated by using the formula:

Salt Index = (Total Na^+ - 24.5) – (Total Ca^{++} - Ca^{++} in $CaCO_3$ x 4.85) Where, all quantities being expressed in ppm.

An irrigation water with a positive index is considered unsuitable for irrigation, while the one with negative index is considered suitable. The Salt Index may vary from-24.5 to 0 for all good waters and to any positive value for all bad waters.

4. Kelly's Ratio

A ratio of Na^+ to $Ca^{++} + Mg^{++}$ is calculated on the basis of equivalents of each element. Ratio must not exceed 1 for good water.

5. Cassidy's 'F' Value

'F' value is the ratio of divalent cations ($Ca^{++} + Mg^{++}$) to monovalent cations ($Na^+ + K^+$) expressed as chemical equivalents. Critical F value proposed is 0.5 or less. When this value exceeds 1 or more, much higher salinity can be tolerated.

6. Bicarbonate hazard

The concentration of carbonate and bicarbonate an-ions are important in irrigation water as regards to the concentration of calcium and magnesium.

The residual sodium carbonate (RSC) is expressed in me L^{-1} as

$$RSC\ (me\ L^{-1}) = (CO_3^{2-} + HCO_3^-) - (Ca^{2+} + Mg^{2+})$$

RSC Value (me L^{-1})	Water quality
< 1.25	Water can be used safely
1.25 – 2.5	Water can be used with certain management
> 2.5	Unsuitable for irrigation

7. Soluble Sodium Percentage (SSP)

Soluble sodium percentage is calculated as:

$$SSP = \frac{Na^+ \times 100}{Ca^{2+} + Mg^{2+} + Na^+}$$

Where, all soluble cations are expressed in me L^{-1}

Irrigation water with SSP value of 60 or more are considered as unsuitable.

8. Boron concentration

Irrigation waters are rated as poor or unsuitable if they contain sufficient boron to damage crops. The permissible limits of boron in irrigation water is sated as:

Quality of irrigation water	Boron conc. (ppm)
Excellent to good	0.3-1.0
Good to permissible	0.7-2.0
Doubtful to unsuitable	1.0 – 3.0
Unsuitable	1.2 – 3.8

9. Chloride Concentration

Chloride ion does not adsorbed on the soil complex and has no effect on physical properties of soil. The chloride concentration is calculated as:

$$\text{Chloride concentration (me L}^{-1}) = \frac{Cl^-}{CO_3^{2-} + HCO_3^- + SO_4^{2-} + Cl^- + NO_3^-}$$

Based on the chloride concentration (me/L), the irrigation waters are classified as:

Chloride concentration (me L^{-1})	Water quality
4	Excellent
4-7	Moderately good
7-12	Slightly suitable
12-20	Not suitable for irrigation purpose
> 20	Not suitable for irrigation purpose

10. Magnesium hazard

Magnesium adsorption on soils affects their physical properties. The magnesium adsorption ratio is calculated as:

$$\text{Mg} - \text{adsorption ratio} = \frac{Mg^{2+}}{Ca^{2+} + Mg^{2+}}$$

A harmful effect on soils appears when the ratio declines below 50.

The United States Salinity Laboratory has proposed a criteria for irrigation water quality (Fig. 8.1). Sodium-adsorption ratio values are used to separate water classes according to hazard from sodium accumulation. A logarithmic scale is used for electrical conductivity on hazards from total salt concentration.

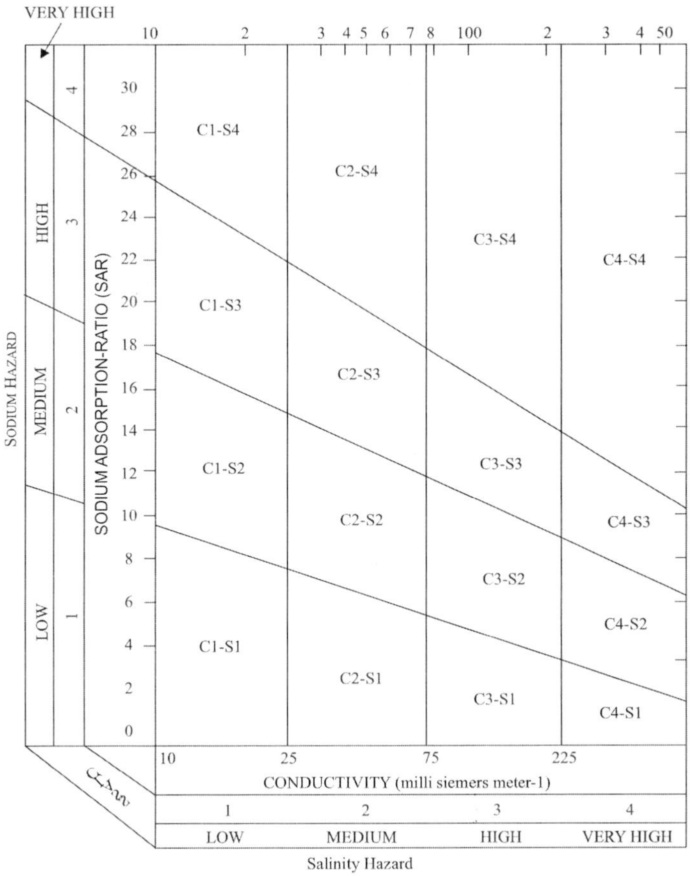

Fig. 8.1: Diagram for determining the quality rating of an irrigation water from its sodium absorption ratio and electrical conductivity. (*Source:* USDA Hand book No. 60, 1954, P. 80)

An irrigation water, to be suitable for irrigation purpose should confirm the following criteria

1. Carbonate should be absent.
2. Amount of bicarbonate expressed as $NaHCO_3$ should not exceed 700 ppm. Water with RSC of more than 2.5 me L^{-1} is not suitable for irrigation. Less than 1.25 meL^{-1} is safe for irrigation
3. Irrigation water having boron concentration of more than 1 ppm is toxic.

   Usually in irrigation water 0.3 to 0.4 ppm B is safe.
4. Amount of total soluble salts should not exceed 1000 ppm. Water with EC of 0.25 dS m^{-1} is safe for use.
5. Sodium should not form more than 75% of the total soluble salts.

Water of high salinity can be used if the amount of soluble Ca is high. In case of high saline or alkaline soils, water with high salt content can also be used without affecting soil structure. Use of water with less salt content may deteriorate the structure due to dispersion which were kept flocculated by the salts.

In addition to the quality of the irrigation water, it is also important to know the nature of soil. Soils in arid regions with heavy texture and high pH, usually develop salinity or alkalinity more quickly than the light textured sandy soils. Besides texture, permeability, drainage, depth of water table, calcium status and pH are the other factors which governs the effect of irrigation water on soil properties.

In areas of good rainfall, marginal water can also be used with advantage as number of irrigation would be small and high rainfall have moderating effect by leaching salts.

Study Questions

1. Explain the following terms
   i) Salinity hazard
   ii) Sodium hazard
   iii) Salt index
   iv) Kelly's ratio
   v) Bicarbonate hazard
   vi) Magnesium hazard
2. Mention different criteria for judging irrigation water quality.
3. An irrigation water contains 460, 150 and 24 mg L^{-1} of $Na^{+}Ca^{2+}$and Mg^{2+} respectively. Calculate :
   i) SAR of irrigation water
   ii) Soluble sodium percentage
   iii) Total cation concentration in meL^{-1}
   iv) Approximate EC in dSm^{-1} and
   v) Osmotic Pressure in bars

9

Properties of Submerged Rice Soils

In India rice is cultivated under two land situations. In uplands, rice is grown under rainfed conditions and the yield is normally low. In medium or low land conditions rice is grown under wet condition where plenty of rain water or irrigation water is available. Here an intensive tillage is practiced in wet to create a soft puddle condition and then paddy is transplanted. Land submergence is thus an essential condition for successful rice cultivation.

Rice soils are not necessarily soils with water at or near the surface. Many soils have a fairly deep water table, but their surface is kept under submergence maintaining depth of 3-5 cm water. This condition is maintained either by puddling when wet or by having a compacted layer in the sub surface.

A rice soil is characterized by reducing conditions, when it is cropped with rice during wet season and oxidizing conditions during dry season, when the land remained fallow or planted with dry season crops if, irrigation water is available.

Effect of Puddling

During puddling, the soil structure is break down, the aggregates lost their identity forming a homogeneous mass of fine and coarse textural separates. Repetition of puddling for rice cultivation year after year leads to formation of hard pan in sub surface horizon.Puddling reduces the non-capillary pore space, micro void increase and apparent volume decreases. It is reported that percolation dominates in sandy soils and seepage in clayeysoil. The rate of percolation and hydraulic conductivity decreases with increase in bulk density.

Characteristics of a typical rice soil profile during growing season

During mid-growing season the soil profile shows following characteristics:

1. There is a film or sheet of water on the soil surface in which sometimes there is growth of algae, which keep this water strongly oxygenated.
2. There is a thin surface layer of oxidized soil, brown or yellowish-brown in color which receives its oxygen from the surface water film. It may be several mm thick.

3. Below this, there is a thick layer of reduced soil, grey or blue – grey in color. The transitional layer between oxidized and reduced layer is fairly sharp, not exactly horizontal which is marked by a thin film of ferric hydroxide.
4. Below this is an oxidized layer again with sharp boundary enriched with ferric hydroxide and layer of black manganese dioxide. This layer remains oxidized during period of flooding because of low microbial activity and of an adequate amount of poisioning material to prevent reducing condition.

The roots of a healthy rice plant remained in this reduced zone containing pockets of soil that are oxidized. Further since fresh organic matter is incorporated during puddling in this layer, the gases produced during formentation of the organic matter escapes upward through the channels, and down ward which oxygenated water can diffuse or percolate. Thus this reduced layer in a rice soil is characterized by a **mosaic of reducing and oxidizing condition**. With progress of growing season, the root system in the reducing layer loses its ability to excrete oxygen, and new root system develops in the thin oxidized surface layer of the soil which remain active till flowering.

Ferrolysis, a model of a cyclic soil forming process in wet-acid soils

The ferrolysis process is responsible for the peculiar properties of the seasonally wet acid soils. In wet season, reduction of ferric oxides produces dissolved ferrous iron, which displaces exchangeable bases. The bases are partially removed by leaching. In dry season, oxidation of exchangeable ferrous iron produces ferric oxides, and exchangeable H+ leading to partial decomposition of clay and releases Al and Si. Ferrolysis is confined mainly to the surface horizons where fresh organic matter is supplied annually.

The process of ferrolysis during wet and dry seasons are depicted below:

Wet Season

Water saturation
↓ → Eluviation of secondary silica
Reduction
↓ → Fe2+ production from free iron oxides
↓
Displacementof Exch.bases and Al by Fe2+
↓
Eluviation of bases, Al and Fe2+

H+ consumption
↓
Formation of soluble Al-Polymers
↓
pH rises
↓
Decrease in Fe2+ activity
↓
Formation of incomplete Al interlayers in 2:1 clay mineral
↓
Trapping of exch. Fe2+ In interlayers

Dry season

Ari entry
↓
Oxidation of exch. Fe2+ to free iron oxides
↓
H+ production
↓
Formation of H+ - clays and pH deceases
↓
Clay decomposition and formation of Al, Mg-clay
↓
Liberation of secondary silica
↓
Formation of secondary silica aggregates

Ferrolysis is not desirable for rice crop growth. This can be managed in several ways such as liming, decrease reduction phase by providing drainage etc.

Changes occurring in a submerged soil

i) Depletion of oxygen

The first change that occurs in a flooded soil is depletion of oxygen. The oxygen diffusion in the water layer above the soil is very slow and the rate of oxygen consumption in reduced layer is high. Because of this high demand of oxygen in flooded soil and slow oxygen supply through water, the soil is practically devoid of oxygen. This rapid depletion of oxygen takes place within a day or so during rice growing season. Some oxygen trapped in tortuous pores is rapidly utilized by facultative anaerobic organisms.

Under anaerobic condition, the micro-organisms responsible for decomposition of fresh added organic matter use free oxygen dissolved in the soil water much faster than atmospheric oxygen can diffuse in the wet soil. In absence of free oxygen, the micro-organism derived their energy through the process of a series of chemical reactions as discussed below.

Under aerobic condition, oxygen accepts electrons and combines with hydrogen ions to give water as per the following reaction:

$O_2 + 4H^+ + 4e^- \rightarrow 2H_2O$

In absence of free oxygen in submerged soils, other substances can accept electrons and lose their oxygen as:

$NO_3^- + 2H^+ + 2e- \rightarrow NO_2^- + H_2O$ and

$2NO_2^- + 8H^+ + 6e^- \rightarrow N_2 + 4H_2O$

Some high valency cations will accept electrons and reduced to a lower valency state and finally the hydrogen ion itself can accept an electron to become hydrogen gas: $2H^+ + 2e^- \rightarrow H_2$

ii) Aeration status of submerged soil

Soil aeration refers to the exchanges of oxygen and carbon dioxide between soil and atmosphere. Carbon dioxide is produced and oxygen is consumed in soil through respiration of plant roots and microbial activity. Soil aeration therefore consists of exchanging O_2 from the atmosphere for CO_2 in the gaseous phase of soil. The rate of diffusion of gases in the air phase is generally greater than in the water phase. Hence, soil aeration is dependent largely upon the volume fraction of air filled pores. Impeded aeration resulting from water logging or soil compaction can strongly inhibit crop growth. The root systems are commonly restricted with decrease in aeration capacity of soil. Poor aeration can decrease uptake of water and induce early wilting.

The volume fraction of air (fa) is related with soil water content as; $fa = f - \theta$ where, f is the total porosity and θ, the volume fraction of water. The volume fraction of air in soil decreases with increasing the water content. The soil aeration is likely to become limiting to root respiration when the volume fraction of air (fa) at field capacity is about 10%.

The exchange of air between soil and atmosphere also occurs by convection and diffusion mechanism. However, the diffusion mechanism is more important for soil aeration than convection. The diffusion mechanism is described by Fick's law of diffusion as, $q_d = - D\ dc/dx$

Where q_d is the diffusive flux, D is the diffusion co-efficient and dc/dx is the concentration gradient.

The diffusion co-efficient in the soil Ds is smaller than that in bulk air Do.

The relationship between Ds and fa is given by:

$Ds / Do = 0.66\ fa$ where,

0.66 is tortuosity co-efficient.

The overall rate of respiration due to all biological activity in the soil i.e. amount of oxygen consumed and amount of carbon dioxide produced by the entire profile determines the aeration requirement of soil.

The aeration requirement of soil depends on several other factors such as temperature, soil wetness, organic matter, types of micro-organism and season of the year.

Anaerobiosis or oxygen stress will occur in soil when the rate of supply fall below the demand. Water logging could severely damage the crop at its most active growth stage.

iii) Accumulation of carbon dioxide

Water logging not only restricts the oxygen supply but also stops the escape of gases into atmosphere. Soil gases like $CO_2$and CH_4 accumulate and may escape as bubbles if pressure is build up. It is reported that during first three weeks of flooding some soils may generate up to 2.5 tonnes of CO_2 per hectare. When the concentration of O_2 in aerating gas is maintained at 21%, growth was not affected by CO_2 concentration as high as 15%. But, the growth of root and shoot reduced when the concentration of CO_2 increased to 45 to 60%. The relation between pH and CO_2 concentration in ferruginous soil is given by: $pH = 6.45 - 0.33 \log p(CO_2)$ where,

$p(CO_2)$ is the pressure of CO_2 in bars. If the soil contains free calcium carbonate, the relationship between the soil pH and the CO_2concentration is:

$pH = 6.03 - 0.67 \log p(CO_2)$

The partial pressure of CO_2 will determine the solubility of Ca^{2+}, Mg^{2+}, Fe^{2+} and Mn^{2+} in soils. The kinetics of pCO_2 varies markedly with soil properties. In slightly alkaline and neutral soils, low in organic matter attained maximum pCO_2 (0.2-0.3 atmosphere) in 4 to 6 weeks after flooding and declined gradually thereafter. In slightly acid soils, low to medium in organic matter but, rich in Fe and Mn attained peak pCO_2 (0.2-0.4 atmosphere) in 2-3 weeks after flooding. On the other hand in strongly latosols rich in Fe but, low in Mn, a peak partial pressure of CO_2 (0.6-0.8 atmosphere) is attained within 3 weeks of flooding.

The increase in pCO_2 on flooding is due to accumulation of CO_2 produced by aerobic respiration at beginning of flooding followed by anaerobic decomposition of organic matter. The higher pCO_2 in acid soils than in neutral soils can be explained by the presence of higher proportion of H_2CO_3 according to the equation:

$$\frac{H_2CO_3}{(H\,CO_3^-)} = \frac{(H^+)}{K}$$

The decline in pCO_2 after one to four weeks of flooding may be due to:

i) Loss of CO_2 from soil

ii) Dilution effect of CH_4 produced due to organic matter decomposition

iii) Bacterial reduction of CO to CH_2

iv) Removal of CO_2 as insoluble carbonates.

Rapid decline of pCO_2 in soils rich in Fe and Mn may occurs due to chemical process as:

$Fe^2 + CO_3^{2-} \rightarrow Fe\,CO_3 Ks = 10^{-10.3}$

$Mn^{2+} + CO^{2-}{}_3 \rightarrow MnCO_3 Ks = 10^{10.1}$

Effects on rice

i) CO_2 injury may be more widespread

ii) In soil low in Fe and Mn (muck soils) the organic matter may aggravate this injury.

iii) Planting should be delayed for at least two weeks after flooding to minimize CO_2 injury.

Electro – Chemical changes

The main electro – chemical changes in submerged soils are:

1. Change in soil pH
2. Decrease in redox potential (Eh)
3. Increase in specific conductance

1. Soil pH

The pH of most soils tends towards neutral after submergence. Under submergence, the depletion of oxygen, accumulation of CO_2 and onset of reduction process greatly influence the soil pH. On submergence the pH of acid soils increases and alkali soil decreases.

In acid soils, reduction of ferric and manganic compounds and release of OH ions, production of ammonia and CO_2 resulted in pH rise. In calcareous soils flooding produces the reserve effect. At equilibrium, the pH of a submerged soil buffered at around neutrality (6.7–7.5). The pH buffering action of submerged soils is due to the iron and manganese redox system and carbonic acid. A majority of the oxidation–reduction reactions involve either production or consumption of $H+/OH^-$ ions. Microbial decomposition of organic matter in alkali soil produces CO_2 which reacts with water to form carbonic acid. The carbonic acid dissociates into H^+ and HCO_3^- ions. The H^+ ion leads to lowering pH in alkali soils.

In iron-rich acid soils, pH is a linear function of pCO^2. The pH of such soils are apparently determined by metastable substances, $Fe_3(OH)_8$ according to the equation.

$$pH = 6.33 - 1/3 \log p\ CO_2 (Ks = 10^{-17.2})$$

The stable pH attained after many weeks is determined by the solubility of $Fe_3(OH)_8$ and partial pressure of CO_2.

For alkaline soils the relation between pH and pCO_2 is given by the equation:

$pH = pH_0 - \log p\ CO_2$ where, pH_0 is the pH of the equilibrium solution at a pCO_2 of 1.0 atmosphere.

Addition of organic matter hastens the reduction process through liberation of CO_2, production of organic acids, which operates as an electron donor in the oxidation – reduction system. These electrons reduce the inorganic substances such as NO^-_3, SO^{2-}_4 etc.

2. Redox potential

Redox potential measurements are used an index of the oxidation – reduction status of flooded soil. The oxidized or reduced state of chemical elements in the soil gives an idea about its aeration status. In well aerated soil, the oxidized states of elements such as ferric (Fe^{3+}), manganic (Mn^{4+}), nitrate (NO^-_3) or sulphate (SO^{2-}_4) dominate. In poorly waterlogged soils the reduced states of these elements such as ferrous iron (Fe^{2+}), manganous manganese (Mn^{2+}) , oxides of nitrogen (NO_x) and sulphide (S^{2-}) are found.

The oxidation-reduction (redox) potential (Eh) of a soil can be measured by inserting a standard electrode and platinum electrode in soil solution. It will provide a measure of the tendency of a system to reduce or oxidize substances. The value of Eh can be interpreted in terms of the ratios of oxidizing and reducing materials present in the soil system. Under strong oxidizing condition, the Eh value is positive and decreases with increasing percentage of saturation. Under anaerobic water logged paddy soils, it is negative.

The oxidized and reduced forms of certain elements with their Eh values are given in table 9.1 (Patrick and Reddy, 1978).

Table 9.1. The oxidized and reduced state of certain elements in soil and Eh at which change in form occurs.

State of element		Eh (volt)
Oxidised	**Reduced**	
O_2	H_2O	0.38 to 0.32
NO_3^-	N_2	0.28 to 0.22
Mn4+	Mn2+	0.28 to 0.22
Fe3+	Fe2+	0.18 to 0.15
$SO2-_4$	S2-	- 0.12 to -0.18
CO_2	CH_4	- 0.20 to -0.28

Flooding or water logging of the soil has a marked effect on the behavior of several plant nutrients. These chemical transformations are almost set in motion by biological (oxidation – reduction) process that result from oxygen depletion.

In the process of microbial respiration in waterlogged soil, iron serve as an electron accepter whereas, oxygen full fills this function in aerated soils.

The redox potential of aerated soils ranged from 500 to 700 mv and decreased to -250 to -350 mv under submergence depending on the degree of saturation, type of organic matter, Fe and Mn content, pH and temperature. The effect of water saturation on Eh is given below:

Kind of soil	Eh (mv)
Aerated soil	+700 to +500
Moderately reduced	+400 to +200
Reduced	+100 to -100
Highly reduced	-100 to -300

On flooding the anaerobic organisms utilized the oxidized soil components as electron acceptor resulting into reduction of these compounds. Organic matter served as an electron donor and enhanced the reduction process.

Nitrate and manganese dioxide are reduced at high potentials, while at extremely low potential sulphate may be reduced. Thus, on submergence due to depletion of oxygen, the reduction of oxidized soil components takes place in a sequential form as listed in table 9.2 and table 9.3.

Table 9.2. Oxidatin-reduction potentials of typical soil systems.

System	Oxidation-reduction potential at 25°C(mv)	
	At pH 5.0	At pH 7.0
$O_2 + 4H^+ + 4e^- = 2H_2O$	930	820
$Eh = 1.23 + 0.0148 \log pCO_2 - 0.059$ pH		
$NO^-_3 + 2H+ + 2e- = NO-_2 + H_2O$	530	420
$Eh = 0.83 - 0.0295 \log NO^-_2/NO^-_3 - 0.059$ pH		
$Fe(OH)_3 + 3H^+ + e^- = Fe^{2+} + 3H_2O$	170	-180
$Eh = 1.06 - 0.059 \log Fe^{2+} - 0.177$ pH		
$SO^{2-}_4 + 10H^+ + 8e^- = H_2S + 4H_2O$	-70	-220
$Eh = 0.30 - 0.0074 \log H_2S/SO^{2-}_4 - 0.074$ pH		
$CO_2 + 8H^+ + 8e^- = CH_4 + 2H_2O$	-120	-240
$Eh = 0.17 - 0.059 \log pCH_4 / pCO_2 - 0.059$ pH		
$2H^+ + 2e^- = H_2 Eh = 0.00 - 0.59$ pH	-295	-413

Source: Russel, E.W.(1973) Soil conditions and plant growth. 10 th Edition, pp. 670-695.

The redox system in flooded soils cover a wide potential range from oxygen system to the hydrogen system, but their identification and characterization is difficult.

The reduce situation in rice culture gives three district features depending on the soil and water management practices.

1. In uplands due to highly oxidized state, NH_4-N is oxidized to NO^-_3-N which may leach downward with percolating water or denitrify. Iron in Fe^{3+} state not in available form. The availability of P is limited. In low pH, aluminium toxicity may arise.

2. In moderately reduced state, NH_4-N may dominate and adsorbed on clay complex, not lost through leaching or denitrification. Ferric iron is reduced to Fe^{2+} and increased its availability. Phosphorus and Mn availability also increases. The soil pH is around 6.5 at which the availability of major nutrients increases.

3. Highly reduced toxic state occurs due to prolonged submergence. Under these situations the rice plants suffers due to iron toxicity. Rice roots are affected due to H_2S toxicity.

Table 9.3. Critical redox potential values of some important oxidizing components in submerged rice soils.

Reduction component	Redox potential (mv)
$O_2 \rightarrow H_2O$	380 to 320
$NO^-_3 \rightarrow N_2$; $Mn^{4+} \rightarrow Mn^{2+}$	280 to 220
$Fe^{3+} \rightarrow Fe^{2+}$	180 to 150
$SO^{2-}_4 \rightarrow S^-_2$	-120 to -180
$CO_2 \rightarrow CH_4$	-200 to -280

3. Specific Conductance

The specific conductance of a solution is a measure of its ionic content. On flooding , the specific conductance of soil increases as Ca^{2+}, Mg^{2+} are mobilized by CO_2 and cation exchange, Fe^{2+} and Mn^{2+} go into soil solution following reduction and NH_4^+ accumulation.

In slightly alkaline soils specific conductance increases entirely by Ca^{2+} and Mg^{2+}. Organic matter enhances the solubility of Ca and Mg and increases specific conductance.

In acid soils, specific conductance increases because of Fe^{2+} while Mn^{2+} and NH_4^+ apparently do not.

Nutrient transformation in submerged soils

Nitrogen transformation

The nitrogen supply to rice under submergence is available from three sources:

i) NH_4-N present in soil when it is flooded.

ii) Mineralization of N from organic matter and crop residues under waterlogged condition and

iii) N added as fertilizer.

Nitrate and nitrite nitrogen present in soil at the time of submergence are not available to crops because of loss through denitrification.

Mineralization of nitrogen

Mineralization of nitrogen is limited to ammonification stage under waterlogging due to depletion of oxygen which is essential for nitrification. Nitrogen mineralization is also slow under submergence as compared to aerated soils because of low efficiency of anaerobic organisms. Incomplete decomposition resulted in production of methane, organic acid which are injurious to rice crop. Wider C:N ratio under anaerobic condition observed due to low N requirement of microbes.

Accumulation of NH_4- N

In flooded soils, mineralization of organic N stops at the ammonium stage since absence of oxygen inhibit the activity of Nitrosomonas and Nitrobactor that oxidizes NH_4^+ into NO^-_2 and NO^-_3 .

Organic – N Mineralization $\rightarrow$ NH4+ (stops at this stage)

Therefore accumulation of NH^+_4 -N in flooded soil is considered as a good index of nitrogen supply to rice crop.

The kinetics for total NH_4-N released during flooding can be described by the equation:

$Y = a + b \log x$ where,

Y, is the total NH_4 – N as ppm in dry soil, x is the time of submergence in weeks. The parameter 'a' is the NH_4^-N content of the soil one day after flooding. The parameter b is highly correlated with the nitrogen and organic matter content of the soil and is a good measure of the velocity of NH_4^-N release in a flooded soil.

The release of NH^+_4 and NO^-_3-N is influenced by the percentage of water content in soil. The data presented in table 9.4 showed that the production of NO^-_3 N occurs under aerobic condition and sharply decreased with increase in water supply. However, production of NH_4 – N occurs both under aerobic and anaerobic conditions.

Table 9.4. Production of ammonium and nitrate in soils incubated at different percentage of water holding capacity.

Percentage of water holding capacity	Relative quantities of mineral N produced during incubation	
	Ammonium-N	Nitrate- N
10	2	11
20	9	17
30	32	31
40	67	62
50	81	86
60	100	100
70	78	40
80	56	10
90	48	-
100	44	-

Addition of organic residues build up high concentration NH^+_4 – N from the beginning of flooding. Addition of paddy straw, grasses, stubbles with wider C: N ratio may depress the ammonium availability, especially in soils with low organic matter.

Loss of Nitrogen

In waterlogged soils, nitrogen is lost through several processes, such as denitrification, volatilization, fixation, leaching and run off losses of nitrate and ammonia.

i) Denitrification

Denitrification is a type of nitrate reduction in which the nitrogenous products are gases. The principal products are elemental nitrogen (N_2) and nitrous oxide (N_2O), but nitric oxide (NO) is of relatively minor importance. On water logging when the oxygen concentration in a soil falls below 0.1 ppm in soil solution, facultative anaerobic micro-organisms used oxidized form of N as electron acceptors in their respiration and reduce them to N_2 or N_2O. Nitrate being in a high state of oxidation, and soluble in water used by facultative microorganisms and it disappears immediately.

Under favourable conditions as much as 1400 kg/ha NO_3–N may be lost within a week of flooding a soil. Neutral soils well supplied with organic matter may lose appreciable amounts of nitrogen on flooding. In acid soils the loss is less because of slow nitrification in the aerobic state.

In flooded soils, special precautions should be taken in fertilizer use. If ammonium fertilizers are applied to oxidized zone, the nitrate formed may leach to reducing zone and denitrify. Sub-surface placement of ammonium nitrogen should be practiced.

ii) Ammonia volatilization

Volatilization loss of N as NH_3 gas occurs when the urea fertilizer is applied on soil surface instead of being incorporated into soil. The volatilization loss increases under high temperature and high pH.It was reported that under laboratory conditions at room temperature about 87% of applied ammonium sulphate was volatized. But the losses are much less under field condition. Under puddled condition, the loss of N through volatilization from a rice field in Bhubaneswar (Odisha) was about 5% when urea was applied. But it was reduced to 2-3% when USG (urea super granule) was placed at 5 cm soil depth. The volatilization loss was higher in summer rice than that in kharif rice. Peak volatilization loss occurs between 4-6 weeks of fertilizer application.

The lower volatilization loss of NH_3-N from USG is attributed to slower rate of upward migration of NH_2-N into the flood water.

iii) Leaching loss

Under prolonged submergences in rice culture, nitrate leach out of the root zone and get denitrified. Intensity of leaching depends on the rate of percolation. In coarse-textured soils with high hydraulic conductivity considerable leaching loss occurs. Leaching and denitrifiction of NO_3constitute the major loss of N in water logged soils.

Phosphorus Transformation

Soil submergence influences the transformation and availability of both native and applied phosphorus. The transformation of phosphorus is linked with pH and Eh changes in submerged soils. The increase in solubility of water soluble P on flooding a soil is attributed to a number of reactions. Increase in P under water logging is due to the hydrolysis of $AlPO_4$ and reduction of $FePO_4$. Soils high in organic matter and low in active Fe showed moderate increase in available P. The slightly acid soil, medium in organic matter and Fe showed no marked increase in water soluble P as a result of flooding.

The magnitude of water soluble P increased with decrease in redox potential. It was reported that the extractable P increased from 10 to 35 ppm between the potential of +200 to -200 mv. Increase in solubility of calcium phosphate in calcareous soils associated with decrease in pH caused by accumulation of CO_2.

Addition of organic matter through crop residues or green manuring may affect phosphate transformation in waterlogged soils through the mechanism of reduction and chelation. Both these processes increase the solubility and availability of P in soil. Alternate wetting and drying reduces the availability of Al-P fraction and increase Fe-P fraction.

Potassium Availability

The low availability of potassium in submerged soil associated with poor aeration may be due to a decrease in the tendency of the plant to absorb potassium rather than a change in chemical status of potassium in soil. The physiological effect of water logging on availability seems to be more pronounced with potassium than with any of the other nutrients.

The results of an experiment of Lowton(1946) on corn illustrate the contrast behavior between potassium and certain other nutrients (Table 9.5).

Table 9.5. Relative yield of nutrients and dry matter yield of corn grown under aerated and nonaerated condition in a silt loam soil.

Nutrient	Relative yield (non aerated/aerated)
Potassium	0.3
Nitrogen	0.7
Phosphorus	1.3
Calcium	0.9
Magnesium	0.8
Dry matter	0.6

The absorption of potassium by corn plants is reduced either by a deficiency of oxygen or an excess of carbon dioxide. Although the potassium absorption by plants is low, its availability in soil increases on submergence. With flooding, soluble Fe^{2+} and Mn^{2+} ions increases resulting release of K^+ into soil solution.

The increase in soluble K+ after submergence is closely related to the concentration of Fe^{2+} in the soil solution. Continuous submergence and alternate wetting and drying increased the exchangeable K^+ in soil.

Nutrient management of rice in Submerged soil

Under submergence, as a result of depleted oxygen supply, the surface soil profile have two distinct layers:

i) An oxidized or aerobic layer near soil surface.

ii) The reduced or anaerobic layer below the soil surface.

A healthy rice plant has its root systems entirely in the reducing layer surrounded by a brown sheath of ferric hydroxide. The reducing layer in soils less suited to rice typically differs from that in good soil. The reducing layer in the soil have a redox potential sufficiently low for production of hydrogen sulphide. Under impeded drainage condition sufficient concentration of iron may leads to iron toxicity problem in rice which can be ameliorated by liming, application of potassic fertilizers and silicate slag.

The farmer can overcome the severity of unfavourable conditions in a number of ways. Incorporation of crop residues, organic manure or green manure crops should be done at flooding so that active microbial decomposition will take place within first few weeks after flooding. The transplanting of rice should be delayed for two to three weeks after flooding to allow first flush of decomposition to pass so that rice roots will be in a favourable environment from the beginning.

The use of nitrogen fertilizers requires care due to the rapid transition that occur between aerobic and anaerobic layers. In aerobic layer, ammonium is converted into nitrate ion and in anaerobic layer nitrate get denitrified. The ideal method for use at beginning (transplanting) is to place urea or ammonium fertilizer in reduced zone (preferably 5 cm below surface) to get optimum benefit. The nitrate fertilizer should be used for top dressing in midseason.

Rice does not respond to phosphate fertilizer because the phosphate from ferric phosphate get released during submergence with rise in soil pH. However high yielding varieties of rice requires high level of phosphorus. Potassium is often needed for rice and potassium chloride is suitable.

Sulphur nutrition of rice raises problems because application of ferrous sulphate may be oxidized to ferric sulphate in the immediate vicinity of roots. Gypsum is a suitable sulphur source for rice under submergence because calcium in gypsum percolated downward, neutralizes subsurface aluminium toxicity and increase percolation rate. Rice responded well to silicate because it increases strength and rigidity of their cells. It increases oxidizing power at the surface of rice roots. It also alleviates manganese toxicity ensuring a rapid oxidation of manganese within or just outside the roots.

Zinc deficiency in submerged rice soil is very common owing to the combined effect of increased pH, HCO_3^- and sulphide (S^{-2}). Application of zinc sulphate @ 5kg Zn/ha at planting was found optimum for higher rice yield.

References

Lowton, Kirk (1946)The influence of soil aeration on the growth and absorption of nutrients by corn plants. Soil Sci. Soc. America Proc. 10: 263-268.

Patrick, W.H. (Jr.) and Reddy, C.N. (1978) Chemical Changes in Rice Soils. In Soils and Rice.IRRI, Los Banos, Phillippines. pp. 361-379.

Russel, E.W. (1973) Soil Conditions and Plant Growth. 10th Edition. pp. 670-695.

Study Questions

1. Briefly discuss the physical changes that occur due to puddling.
2. Rice soil is characterized by a mosaic of reducing and oxidizing conditions – comments on this.
3. What do you mean by ferrolysis? Does ferrolysis occurs in well drained rice soils?
4. Briefly discuss the chemical changes occur in submerged rice soils.
5. Explain the beneficial and harmful effects of submergence on nitrogen mineralization and uptake by rice.
6. Give a sound nutrient management strategy for rice in submerged soils.

10

Soil Pollution

Pollution

Pollution is an undesirable change in physical, chemical or biological characteristics of air, water or land caused by excessive accumulation of pollutants (substances causing pollution).

Kind of pollution

Pollution can be classified in many ways:

1. On the basis of part of environment where it occurs most: Air pollution, water pollution and soil pollution.
2. On the basis of origin, pollution may be

   Natural-Volcanic eruption add tonnes of toxic gases and particulate matter in the environment.

   Anthropogenic-Pollution caused due to human activities such as industrial pollution, agricultural pollution etc.
3. According to the physical nature of the pollutants, the pollution named as gaseous pollution, dust pollution, thermal pollution, noise pollution, radioactive pollution.
4. From the ecosystem viewpoints:

   Pollutants can be categorized as: **Non-biodegradable** and **Biodegradable pollutants**.

Non-biodegradable materials such as DDT, BHC, plastic bottles, polythene bags etc. are either not degraded or degraded slowly by decomposers in the nature.

Biodegradable pollutants such as market garbage, lives stock wastes, municipal sewage etc. can be decomposed efficiently by the decomposers. If, managed properly, biodegradable wastes can be turned into useful resources.

Soil Pollution

The soil is a primary recipient of waste products and chemicals used in modern societies. It has been convenient to 'throw things away" and the soil has been the recipient of most of these things. Every year million tonnes of products from a variety of sources-industrial, domestic, solid wastes from cities, heavy metals, solvent find their way into the soils. Once these materials enter into soils, they become part of biological cycles that affect all forms of life. Although, the soil can act as a physical filter by its sieving action, a chemical filter by adsorbing and precipitation of chemical substances, and a biological filter by decomposing organic materials, it does not have infinite capacity to perform these functions. The production of greenhouse gases such as nitrous oxide, methane and carbon dioxide, sulphur and nitrogen containing gases coming from industrial and domestic sources acidify the atmosphere and come to earth as acid rain. Many pollutants reaching the soil can build up to such concentration which pose serious problems to plants and animal health. Impact of soil pollution is not immediately visible but shows its effects only after many years.

Air pollution and their effects on soil and plants

Degradation of air quality and natural atmospheric conditions constitute air pollution. An air pollutants may be a gas or particulate matter i.e. suspended aerosols composed of solids and liquids.

Natural sources of air pollution include pollen, dust and smoke from forest fires and volcanic ash.

Anthropogenic air pollutants enter the atmosphere from fixed sources (factories, power plants, mineral smelters), while mobile sources include all sort of transport vehicles moving by road, rail or air.

Classification of air pollutants

Air pollutants can be classified into two categories viz. **(1) Primary and (2) secondary air pollutants.**

1. Primary air pollutants and their effects

Primary pollutants enter the atmosphere directly from various sources. These includes particulate matter, carbon monoxide (CO), hydrocarbons (HCs), sulphur dioxide (SO_2) and nitrogen oxides (NO_x).

Particulate Matter

Particulate matter comprises of solid particles or liquid droplets (aerosols) like smoke, dust, asbestos fiber, pesticides, some metals (Hg, Pb, Cu and Fe) which remain suspended in air. The particles having diameter more than 10 micrometer generally settle on plant and soil surface in less than a day. The particles less than 1 micrometer can remain in air for a week or long.

Carbon monoxide (CO)

It is a product of incomplete combustion of fossil fuels. It is short lived in the atmosphere and gets oxidized to CO_2. Carbon monoxide is highly poisonous to animals. When inhaled, CO reduces the oxygen carrying capacity of blood.

Hydrocarbons (HCs)

Hydrocarbons are compounds composed of hydrogen and carbon. Hydrocarbons are produced naturally during decomposition of organic matter. Methane (CH_4) is evolved from soil in flooded rice fields and swamps. Hydrocarbons are also generated during burning of fossil fuels.

Sulphur dioxide (SO_2)

Sulphur dioxide is produced when sulphur containing coal is burnt. Ore smelters and oil refineries also emit significant amounts of SO_2. Exposure of plants to high SO_2 concentration is harmful.

Nitrogen oxides (NO_x)

Nitrogen oxides are formed from N_2 and O_2 during combustion of fossil fuels at high temperature in automobile engines. Nitrogen oxides contributes to heart and lung problems and also contribute to acid rain.

2. Secondary air pollutants and their effects

Secondary pollutants are formed during chemical reactions between primary air pollutants and other atmospheric constituents. Generally these reactions occur in presence of sunlight.

Photochemical smog is a classical example which is formed in traffic-congested cities where warm conditions and intense solar radiation are present.

Photochemical smog is mainly composed of ozone (O_3), peroxyacetyl nitrate (PAN) and NOx. Automobile exhaust containing HC and NO play an important role in O_3 and PAN formation. The photochemical reactions involved in smog formation is as follows:

Reaction occurring in engine

$N_2 + O_2 \rightarrow 2\ NO$

Reaction occurring in atmosphere

$2NO + O_2 \rightarrow 2\ NO_2$

$NO_2 \xrightarrow{\text{UV radiation}} NO + O$

$O+O_2 \rightarrow O_3$

$NO + O_3 \rightarrow NO_2 + O_2$

$HC + NO + O_2 \rightarrow NO_2 + PAN$

Smog ozone may damage plant as well animal life. Ozone aggravate lung diseases in humans. Several plant species are very susceptible to PAN in smog. PAN damages chloroplasts and thus reduces plant growth.

Acid Rain

Acid rain or acid deposition results when gaseous emissions of acidic oxides like CO/CO_2, SO_2/SO_3 and NO_x interact with water vapour and sunlight and chemically convert to acidic compounds like H_2SO_4 / H_2SO_3, HNO_3 / HNO_2, H_2CO_3 etc. There may be dry deposition of acid salts. Acid rain is a major global air pollution problem.

About half of the acidity in the atmosphere is transferred to earth through dry deposition. Dry deposited gasses and particles can also be washed from trees by rainfalls. Normally, rainwater is slightly acidic (pH 5.6 – 6.5) because water and CO_2 combined in air to form a weak acid.

Effects of acid rain

Acid rain damages building, monuments, marble structure. Acid rain adversely affects terrestrial and aquatic vegetation. Acid rain increases soil acidity, damages forest ecosystem. Crop damage is a common feature because of acid rain.

Water pollution: Effects on soil and crops

Based on their origin, the sources of water pollutants can be broadly divided into:

i) **Point source**, where the effluent discharge occurs at a specific site. For example- sewage outlet of municipal area or effluent outlet of a factory.

ii) **Non-Point source**, where inflow of pollutants occurs over a large area. For example- agricultural runoff.

Point source pollution can be effectively checked with appropriate technology whereas, non – point source is difficult to control and needs control measures in a large scale.

Water pollutants can be

i) Biological-Pathogens such as bacteria, viruses, protozoa.

ii) Chemical-Organic chemicals and inorganic chemicals like nitrate, phosphate, chloride, heavy metals like As, Pb, Cr, Cd, Hg etc.

iii) Physical-hot water from industries, oil spills from oil carriers.

These pollutants are generated by different sources and activities which are described below:

i) Municipal waste water

Liquid wastes carrying organic and inorganic materials from domestic activities such as kitchen, toilet are discharged directly into a river or large water bodies nearby. Many rivers in India are polluted by indiscriminate discharge of waste waters.

ii) Industrial waste water

Both small scale and large scale industries produce waste waters contaminated with a variety of organic and inorganic pollutants. All most all the rivers of India are heavily – polluted by the discharge of industrial waste water.

iii) Surface run off from agricultural land

The run off from agricultural land is contaminated with pesticide residues and residue of inorganic fertilizers. All these pollutants in runoff heavily contaminate our surface water and ground water resources.

iv) Oil spills

An accidental discharge of petroleum in ocean or estuaries due to off shore oil mining and explore operations contribute to oil pollution of marine ecosystem. The death of plankton, fish, marinebirds is a significant ecological effect of oil spills.

Effects of water pollution

Water pollutants adversely affects the physical, chemical and biological characteristics of the aquatic ecosystem and quality of ground water.

i) Effects on aquatic ecosystem

Organic and inorganic pollutants decrease the dissolve O_2(DO) content of water bodies. Water having DO content below 8 mg L^{-1} is considered as contaminated.

Heavily polluted waters have DO content below 4 mg L^{-1}. The photosynthetic activity, O_2 consumed by organisms and decomposition of organic matter are affected due to low DO content in water. It also affects the biochemical oxygen demand (BOD) and chemical oxygen demand (COD) of water.

ii) Biological Magnification

The phenomenon through which pollutants get accumulated in tissues in increasing concentration along the food chain is called **biological magnification.** Such pollutants are non-biodegradable. Once they are absorbed by an organism, they cannot be metabolized and broken down or excreted out. These pollutants generally get accumulated in fat-containing tissues of the organism.

The classical example of biological magnification is DDT. DDT is sprayed in water bodies to check the growth of mosquitoes. Later, it was found that the concentration of DDT had increased in by 800 times in phytoplankton relative to the concentration in water. In different fish, the DDT concentration was 9 to 40 times higher than that in zooplankton and in birds 25 times higher than that in fish. Many other persistent pesticides and radio nuclides also show biological magnification.

iii) Eutrophication

Eutrophication is a process in which nutrients are accumulated in water bodies leading to loss of biodiversity. Excess nutrients accumulation in water bodies causes profuse growth of algal which covers the entire water bodies and cause deficiency of O_2 in water. Thus in infected water bodies the growth of other species may be inhibited due to lack of O_2 or due to accumulation of toxic materials.

iv) Effect on human health

The sewage water from domestic and industrial areas contains pathogens like bacteria, virus and heavy metals which cause health problems. **Minamata** disease due to consumption of fish contaminated with mercury was detected in Japan in 1952. Cadmium pollution can cause **Itai-Itai**disease that effects liver and lung. Excess nitrate in drinking water may be fatal for infants. The **blue-baby syndrome** in infants is caused due to excess nitrate intake through drinking water. Excess fluoride in drinking water cause **fluorosis** in both animals and human body. Excess concentration of arsenic in groundwater may cause **black – foot disease.** Arsenic may also cause diarrhea, lung and skin cancers.

Contaminated soils

In certain areas, the soils are getting contaminated with toxic elements either from geogenic sources (Arsenic, Selenium, Fluoride) or from application of sewage water, industrial effluents, solid wastes, pesticides, fertilizer and radionuclides. The pollutants having entered into food chain become potential human, animal health hazard.

1. Arsenic toxicity

Arsenic builds up in soil environment through natural process of weathering of arsenic bearing rocks, or use of arsenic contaminated ground water for irrigation or through anthropogenic activities such as mining, smelting of metals, combustionof coal and application of arsenical pesticides. Out of 20 countries (Argentina, chille, Finland, Hungry, Mexico Nepal, Taiwan, Bangladesh, India etc.) in the world, the magnitude is highest in Bangladesh followed by India. The widespread arsenic contamination in groundwater above the permissible limit (50 mgL^{-1}) has been detected in several places of the country (Table 10. 1).

Table 10.1. Arsenic contaminated ground water in different states of India.

State	No. of districts affected	Level of contamination (micro gram L^{-1})
West Bengal	12	50-3700
Assam	18	> 50
Bihar	12	> 50
Jharkhand	1	> 50
Uttar Pradesh	21	> 50
Madhya Pradesh	1	52-88
Manipur	1	798 – 986
Tripura	3	65 - 444
Nagaland	2	> 50

Natural abundance

Dissolved arsenic concentrations in natural water are generally low. The sedimentary rocks generally have higher arsenic content than the igneous and metamorphic rocks. The suspended and bottom sediments in – aquatic systems contain more arsenic than most of the natural waters. The capacity to retain arsenic is influenced by the sediment grain size and presence of surface coating composed of clays, clay sized iron and manganese oxides and organic matter. Arsenic held by solid phases get released to ground water through oxidation–reduction reaction, ion exchange and chemical transformation.

Arsenic in ground water is generally present as arsenites or arsenate or both, besides the organic forms. The toxicity of arsenic compounds in ground water depends largely on oxidation state and hence redox status and pH. The arsenites are more soluble, mobile and toxic than arsenates in aquatic and soil environment. The biological availability and phytotoxicity of arsenic in soil increases on reduction to As (III) state which is facilitated on flooding of soil. Low land rice seems to be susceptible to As-toxicity.

Health Implication of Arsenic Poisoning

Arsenic is a widely occurring toxic metal in natural ecosystem. As small as 0.1gm of arsenic trioxide is lethal to human. The maximum tolerance limit of in organic arsenic is 0.002 mg As Kg^{-1} body weight for human. Early symptoms of arsenic poisoning include skin disorders, weakness, nausea and vomiting with diarrhea or constipation. Under acute cases the symptoms include skin pigmentation, hyperkeratosis, enlargement of liver, skin cancer. "**Bellville Disease** "in Argentina, "**Black foot Disease**" in Taiwan and "**Kai Dam**" disease in Thailand are well established health disorders due to arsenic poisoning.

Effect on soil and crop

Research results showed that total arsenic content in soils of Nadia and 24pargana districts of West Bengal varies between 8.40 to 24.3 mg As Kg^{-1}soil. The Olsen extractable As varies between 1.22 to 9.80 mg As Kg^{-1} soil. These contents are, generally, higher than those reported for several other countries like, Argentina, China, Mexico, France and Australia. Accumulation of arsenic in crops was more when grown in high contaminated soils. The boro or summer rice using large quantities of underground water contained more arsenic than kharif rice. The leafy and underground vegetables contained much higher arsenic as compared to vegetables with fruits (Table 10.2).

Table 10.2. Arsenic content (mg kg^{-1}) in edible parts of crops grown on arsenic contaminated soils.

Crop	Level of contamination of soil		
	Low (<0.4 mg As Kg^{-1})	**Medium (0.4-0.8 mg As Kg^{-1})**	**High (>0.8 mg As Kg^{-1})**
Rice (Kharif)	0.4-0.5	0.5-1.2	1.6-2.8
Rice (boro)	1.2-1.6	1.9-3.3	4.9
Radish	3.6	4.5-5.1	4.9-6.0
Ladies finger	2.1	2.7	3.8
Beet	-	4.3	4.5-5.1
Knolkhol	-	2.5	2.5-3.1

Source: Sanyal (2001)

Arsenic and phosphorus interaction in soil

Arsenic and phosphorus are both placed in group Vb of the periodic table. The interaction of P and As in soil plant system is an important issues of As mobilization. These two oxyanions would not be absorbed independently in mixtures, but would compete for adsorption on same sites. Several workers have showed that the presence of P caused a reduction in arsenate adsorption by soil minerals.

Remedial Measures

1. Optimum use of ground and surface water. Use harvested rain water during the lean period.
2. Development/ Identification of low arsenic accumulating high yielding crops and varieties for contaminated arsenic areas.
3. Irrigation with pond – stored groundwater – decontamination being facilitated by rainfalls and sedimentation.
4. Increased use of FYM, green manure crops, phosphatic fertilizers, zinc and inclusion of pulses in the cropping sequences are recommended.

2. Selenium toxicity in soil and plant

Selenium (Se), depending upon its concentration may be toxic to plants, animals and humans. Selenium is composed of the enzyme glutathione peroxidase, one of the antioxidant defense system of the body. Selenium deficiency may leads to **Kashin beck** (Chronic arthritis) and **Keshan** (cardiomyopathy) disease. The Food and Nutrition Board of the National Academy of Sciences, USA, has accepted 5 mg Se Kg^{-1} diet as the critical level between toxic and non-toxic feeds. Soils containing more than 0.5 mg Se Kg^{-1}soil are designated as seleniferous soil. The critical limit of Se for drinking water is 10 mg Se L^{-1} and for irrigation water 20 mg Se L^{-1}.

Selenium toxicity problems are seen in throughout the great plains of USA, Prairie regions of Canada and other countries like Australia, China and Punjab state in India.

Selenium content of soils in seleniferous areas in Punjab ranges from 0.85 to 4.52 mg Se Kg^{-1} as compared to normal soils which ranges from 0.23 to 1.08 mg Se Kg^{-1}. Selenium content in soils decreased with depth. In irrigation water it ranges from 2.54 to 69.53 mg L^{-1} in toxic areas where as it is 0.25 to 8.63 mg L^{-1} in nontoxic areas.

Chemistry of Selenium

In neutral or acid soils selenites (Se (IV) forms) is dominant which easily binds to iron to form less soluble iron selenite. In well aerated alkali soils, selenate (Se (VI) forms) is pre dominant and easily taken by the plants. Selenates are more toxic and mobile than the reduced selenites. The retention of Se in soils decreased in the presence of sulphate, nitrate and phosphate ions. Addition of organic matter decreases the availability of Se in soil. Selenium contents in soils increases with fineness of soil texture.

Effects of Selenium Toxicity

The selenium toxicity in soils, ground water, fodder and edible products, is found in the seleniferous region of north-eastern Punjab. The intake of contaminated food and fodder causes chronic **Selenosis** (Selenium poisoning) in animals and humans. In severe cases, there is high mortality of animals besides disorders in the reproductive system. Loss of hair from head, mal formation of finger as well as toe nails health problems are seen in human beings.

Management of Seleniferous Soils

For amelioration of Se-toxic soils, following measures have been recommended.

i) Apply gypsum @ 0.8 to 1 ton ha^{-1} in every alternate year before sowing fodder crops, since it helps in reducing Se absorption by crops by 60-70%.

ii) In Se-toxic areas, Oat should be grown as a fodder crop, because its Se absorption capacity is the lowest among other crops. Avoid feeding the first cut to animals as it contain 2-3 times higher Se than do the following cuts.

iii) The farmers are advised to follow maize – wheat rotation instead of rice – wheat due to low water requirement of the former.

3. Fluoride toxicity in soil-water-plant-animal system

Fluoride is essential for protection against dental caries and weakening of the bones but excess fluoride may lead to dental or skeletal **fluorosis**. However, a minimal level of 0.8 mg F L^{-1} in drinking water is required to effectively protect the dental caries.

Source of fluoride

i) Minerals

The origin of F has been associated with volcanic and plutonic activities in the crust of earth. The primary fluoride containing minerals are: Fluorspar ($Ca\ F_2$) (48.5%), Cryolite (3 NaF. $Al\ F_3$) (54.5%) and apatite (3.8%). Reserve of fluoride containing minerals and rocks are seen in Gujarat, Chhattisgarh, Bihar, Andhra Pradesh, Rajasthan, Himachal Pradesh, Madhya Pradesh, West Bengal etc.

ii) Fertilizer industry

Hydrogen fluoride (HF) is evolved during production of phosphoric acid and single super phosphate fertilizer.

$$Ca_{10}(PO_4)_6F_2 + 10H_2SO_4 + 20H_2O \rightarrow 10CaSO_4 + 20H_2O + 2HF + 6H_3PO_4$$
$$Ca_{10}(PO_4)_6F_2 + 7H_2SO_4 + 3H_2O \rightarrow 3Ca(H_2PO_4)_2 H_2O + 7CaSO_4 + 2HF$$

iii) Industrial Emission of Fluoride

A number of industrial operations that heat F- containing materials at high temperature eject an enormous amount of F in the environment. The industries like aluminium smelters, steel and iron industry, cement and phosphate fertilizer units, oil refinery, tooth paste and automobile released substantial quantities of F into air. Fluoride emitted in air from the industries exists both in gaseous and particulate forms. Hydrogen fluoride, (HF), silicon tetrafluoride (SiF_4), fluorosilicic acid (H_2SiF_6) and carbon tetrafluoride (CF_4) are common gaseous forms. Particulate emissions of F include sodium fluoride (Na F), fluorspar (Ca F_2) ,cryolite ($Na_2 AlF_6$), aluminium fluoride (AlF_3) and calcium fluorosilicate (Ca Si F_6). Most of the industrial F emissions are air-borne and may contaminate soil, water, vegetation in the industrial vicinity and up to a considerable distance from the source of emission. The content of F in soil, water and plant around Aluminium Smelter Plant NALCO at Angul in Odisha is given in table 10.3.

Table 10.3. Fluoride content in soil, water and plant around Aluminium Smelter Plant NALCO, Odisha.

Distance from centre	F content (mg F kg^{-1}soil)
Soil	
0-5 km radius	0.5 to 6.14
5-10 km radius	0.02 to 5.04
Water	mg L^{-1}
0-5 km radius	0.52- 3.86
5-10 km radius	0.36-1.15
Tube well water	
0-5 km radius	0.50- 1.83
5-10 km radius	0.20-1.16
Plant species	mg kg^{-1} dry weight
Arhar	390
Ber	20- 30
Paddy straw	15- 100
Bean	30-125
Amaranthus	20- 90

Source: Jena *et al.* (2003)

Fluoride Content in Soil, Water and Plant

Soil

In most soils, F is associated with mica and other clay minerals. The F content in soils depends upon soil type, depth, and soil pH. Mineral soils have higher concentration of F (200 – 300 ppm) than organic soils. Heavy textured soils have more content of F than coarse textured soils. In the Yamuna alluvial plains of India, content of F ranged from 43 to 193 ppm in coarse textured soils and 248 to 1500 ppm in heavy textured soils.

Water

Fluoride is present in traces in rain water. Toxic level of F (tolerance limit 1-1.5 ppm) in groundwater was recorded in the areas loaded with F bearing minerals. Rivers flowing over alluvium contain 0.12 to 0.35 ppm F while those flowing on limestone have less than 0.4 ppm. As compared to fresh water, sea water has higher F content (average 1.3 mg F L^{-1}).

Plants

Fluoride content in wheat, leafy vegetables and fruits ranges between 0.1 and 1.4 ppm. Tea leaves are usually rich in F. The F content in branded tea leaves marketed in India varies from 60-112 ppm. F levels (ppm) in some food items in India are: Wheat 4.6, rice 5.9, maize 5.6, bengal gram 6.2, cucumber 4.1, tomato 3.4, brinjal 1.2, ladies finger 4.0, carrot 4.1, potato 2.8, onion 3.7, banana 4.2 , mango 3.7, apple 5.7 (Sengupta and Pal, 1971).

Remedial Measures

The remedial measures for correcting the fluoride toxicity in drinking purposes include dilution with good quality water, fluoride removal by precipitations using suitable coagulants like lime-alum mixture, adsorption on activated carbon or activated alumina, reverse osmosis or ion exchange.

Emitted particulate air pollutants from the aluminium smelter plants can be controlled by using arresters and scrubbers.

4. Toxicities from urban solid wastes (USW)

USW are non-liquid waste materials arising from domestic, trade, commercial activities. More than 450 cities in India generate about 17 x 10^6 m^3 of raw waste per day. With available treatment plants, production of sewage sludge is estimated to be around 1200 tonnes per day. The nutrient potential of available sewage in India is estimated to be more than 3.5 lakh tonnes of N, 1.5 lakh tonnes of P and2.0 lakh tonnes of K per year. Obviously, all the sewage sludge will be applied to soil. Mostly, much of the solid waste is disposed in low lying areas without taking proper precaution. leachate from these areas can pollute ground water. The Composition of USW from different cities in India are given in table 10.4 and 10.5.

USW contain toxic amounts of heavy metals like Zn, Mn, Cd, Cr, Ni, Pb etc. Continuous application of UWS will increase the organic and in organic load beyond the assimilation capacity of soil. The excessive N loading creates imbalance C: N ratio in soil.

In many countries treated USW are commonly used as soil amendments. However, continuous use of such wastes on agricultural land may contaminate soil and ground water.

Table 10.4. Composition of USW from different cities of India.

Composition (%)	Cities					
	Ahmedabad	Delhi	Nagpur	Chennai	Jaipur	Kolkata
N	1.18	1.67	1.08	0.82	2.34	0.56
P	0.56	0.56	0.51	0.65	0.94	0.12
K	0.23	0.11	0.21	0.12	0.12	0.59
Organic matter	33.5	54.7	39.1	40.2	39.1	3.19
Cu (mg kg^{-1})	535	440	272	210	265	-
Zn	2147	1610	832	935	1720	-
Cd	3.5	5.5	1.5	8.3	7.3	-
Cr	60.4	53.5	49.2	38.5	176	-
Ni	32.3	815	14.8	60.5	37.5	-
Pb	76.8	34.5	24.3	16.6	66.9	-

Source: Juwarkar *et al.* (1991)

Table10.5. Physico–chemical properties of USW Bhubaneswar

Properties	Values
pH	6.9-7.5
EC (dS/m)	0.10-1.75
Exchangeable (cmol(+) kg^{-1})	
Ca	6.5-19.05
Mg	0.89-3.2
Na	0.2-0.8
K	0.07-0.21
Organic carbon (%)	1.06-4.8
Available (kg ha^{-1})	
N	505-780
P_2O_5	98-205
K_2O	350-515

Available micro and secondary nutrients (ppm)	
$SO_4 - S$	15.0-56.9
Fe	56.0-250.0
Mn	59.94-194.0
Cu	4.94-19.56
Zn	3.76-14.05
B	0.76-1.52
Mo	2.76-4.0
DTPA extractable heavy metals (mg kg^{-1})	
Pb	2.30-8.35
Cd	Trace – 1.50
Cr	0.13-3.5
CN	0.02-0.80
WSF	0.52-2.13
As	9.53-12.76
Co	0.94-1.97
Hg	1.05-2.64
Ni	1.06-3.55
Se	7.36-24.50

Source: Yerra (2005)

The value of USW after composting is a good source of organic manure and plant nutrients. Its application to agricultural lands is desirable to replenish the soil fertility. The value of compost, however, is questioned because of its threats to soil quality and human health. For proper safeguard on its use, the regulatory authorities must ensure compliance of ISI quality standards for organic fertilizer (compost) by the manufacturers (Table 10.6).

Table 10.6. Maximum permissible limit of heavy metals in compost.

Heavy metals	Conc. not exceeding (mg kg^{-1} dry basis)
As	10.0
Cd	5.0
Cr	50.0
Pb	100.0
Cu	300.0

Source: ISI (2002)

Effects of USW on soil

- Sewage sludge contains appreciable amounts of plant nutrients and heavy metals.
- Application of sludge make a soil fertile and improve physical properties.
- Application of sludge @ 40-80 tonnes ha^{-1} increased heavy metal contents in soils but, @ 10-20 tonnes ha^{-1} had no toxic effects.

- Cumulative accumulation of Cd, Cr, Cu, Ni, Pb and Zn in sewage sludge treated soils are confined to 0-15 cm soil depth.
- Careful management of sewage sludge either directly or after compositing can make an unproductive soil to productive one.

5. Toxic effects of sewage effluents on soil and crop

In India, it was estimated in 2003 that 22,900 million litres per day of domestic waste water generated from urban centres. The effluents, in general, contain appreciable amounts of organic matter and plant nutrients.Generally around the cities, the farmers use sewage water for irrigation purpose. Long term use of these waste water on agricultural lands leads to potential buildup of heavy metals in soils because of their long residence time in soils. Crops raised on metal contaminated soils accumulate metals in toxic amounts excessive enough to cause clinical problem in animals and human beings consuming these metal rich food products.

Several studies from Hyderabad, Kolkata, Ludhiana, Amritsar indicated that the soils irrigated with sewage water for years leads to accumulation of Cu, Fe, Zn, Al, Cr, Ni in toxic amounts. The heavy metal contents in vegetables irrigated with sewage water in Dappa area of Kolkata were 2-40 times higher than that with normal water irrigation.

Yerra(2005) studied the effects of sewage water on soil and crops in central farm, OUAT.Sewage water has been used for irrigation in E-block of central farms of OUAT, Bhubaneswar since more than 30 years.The sewage water is neutral in reaction (pH 6.89) and rich in plant nutrients such as NH_4-N, Ca, Mg, K and SO_4-S. Toxic amounts of Pb (4.3 ppm), As (6 ppm), Hg (4.3 ppm) make the water unsuitable for irrigation. Selenium was absent.

The soils contain toxic amounts of Fe, Mn, Pb, Cr, Ni, As, Hg and Se (Table 10.7). Accumulation of Fe, Mn, Pb, Cr and As in toxic amounts were observed in different crops such as okra, tomato, maize, cabbage, cauliflower and sunflower. The leafy vegetables like green amaranthus and cabbage recorded higher accumulation of heavy metals as compared to other vegetables (Table 10.8 and 10.9).

Vertical distribution of heavy metals

Maximum accumulation of heavy metals in contaminated soils is confined to 015 cm soil depth (Table 10.10). Because of the low mobility of heavy metals the content of Fe, Mn, Zn, Cu, Cr, Pb and Cd decreased with soil depth.

Table 10.7. Heavy metal content of soils irrigated with sewage water (Central farm, OUAT, Bhubaneswar).

Parameters	Site-1	Site-2	Site-3	Site-4	Site-5	Normal Soil	Sewage water
Clay (%)	22.5	18.9	25.3	24.0	28.0	17.5	
pH	6.3	6.5	7.0	7.2	7.2	6.2	
O.C (%)	0.74	1.02	0.80	0.88	0.78	0.65	
DTPA Extractable (ppm)							
Fe	122	72	63	77	103	61	1.5
Mn	25	13.4	15.6	62.6	17.0	15.0	0.2
Zn	5	3.7	4.7	5.4	44.7	0.8	2.5
Cu	4	2.4	3.5	6.2	11.6	0.8	0.1
Mo	4	3.8	3.8	3.8	3.8	0.6	1.6
Av.B	0.9	1.0	1.5	0.9	1.2	0.5	0.8
Heavy metals (ppm)							
Pb	9.4	7.8	7.6	7.4	8.6	0.1	4.3
Cd	0.8	0.6	0.6	0.6	0.7	0.04	0.2
Cr	2.1	2.2	3.2	2.8	3.0	0.6	0.03
Ni	3.6	2.0	2.1	3.5	3.8	0.06	0.62
As	13.0	12.0	11.0	12.5	12.5	1.8	6.0
Co	2.6	2.4	2.1	1.8	2.2	0.34	0.09
Hg	3.0	2.6	2.3	2.6	2.5	0.02	4.3
Se	25.0	10.0	10.2	10.6	10.0	0.01	Absent

Source: Yerra (2005)

Table 10.8. Micronutrient content in some crop plants irrigated with sewage water (OUAT, Bhubaneswar).

Crops	Micro nutrient content (mg kg^1)					
	Fe	Mn	Zn	Cu	B	Mo
Okra	680.0	114.0	39.0	5.0	27.5	9.8
Tomato	750.0	133.0	65.0	9.8	52.3	13.2
G. Amaranthus	725.0	914.0	81.0	13.2	15.3	14.2
Brinjal	615.0	219.0	46.0	9.8	16.2	9.5
Rice (summer)	2144.0	416.0	56.6	9.8	22.5	10.0
Rice (kharif)	3088.0	705.0	75.8	9.8	25.6	9.9
Mustard	680.0	153.0	39.0	8.2	33.2	6.1
Maize	951.0	114.0	43.0	11.4	31.6	5.2
Cabbage	725.0	1166.0	171.2	114.5	65.0	13.1
Cauliflower	625.0	424.0	150.1	98.0	58.6	13.0
Sunflower	515.0	323.0	54.8	10.5	36.5	3.5

Source :Yerra (2005)

Table 10.9. Heavy metal content in some crop plants irrigated with sewage water (OUAT, Bhubaneswar).

Crops	Heavy Metal Content (mg kg[1])							
	Cd	Pb	Cr	As	Co	Hg	Ni	S e
Okra	0.32	3.94	3.7	6.3	0.9	1.6	0.1	0.0
Tomato	0.39	4.91	2.9	6.5	1.2	1.6	0.5	0.0
G. Amaranthus	0.56	4.81	4.0	7.6	1.1	1.8	10.9	5.0
Brinjal	0.38	4.71	7.2	7.2	1.8	1.6	0.3	0.0
Rice (summer)	0.49	6.45	3.1	8.4	1.7	1.3	1.3	0.0
Rice (kharif)	0.49	6.45	3.0	8.3	1.6	1.4	1.1	0.0
Mustard	0.38	3.03	2.1	8.2	1.5	1.7	1.9	0.0
Maize	0.50	8.20	6.6	8.9	1.2	1.8	1.6	0.0
Cabbage	0.74	6.1	4.8	8.6	2.1	1.8	2.0	0.0
Cauliflower	0.54	5.5	3.1	7.9	2.2	1.7	1.8	0.0
Sunflower	0.40	4.2	2.5	8.0	2.1	1.2	1.7	0.0

Table 10.10. Vertical distribution of heavy metal in soil irrigated with sewage water (OUAT, Bhubaneswar).

Soil depth (cm)	Clay (%)	O.C (%)	DTPA extractable (mg kg $^{-1}$)						
			Fe	Mn	Zn	Cu	Cr	Pb	Cd
0-15	19.6	0.89	197	91	5.1	5.1	3.2	8.5	0.24
15.30	21.1	0.85	33	58	3.3	3.3	2.1	1.0	0.21
30-45	23.8	0.75	8	14	1.6	1.6	0.9	0.98	0.21
45-60	23.8	0.75	8	8	1.3	1.3	0.8	0.95	0.20
60-75	25.1	0.74	4	4	1.7	1.7	0.7	0.94	0.21
75-90	27.0	0.74	5	4	1.3	1.2	0.1	0.88	0.22
90-105	28.2	0.76	13	7	1.7	1.3	0.09	0.89	0.20

Source :Yerra (2005)

Industrial effluents-soil and water pollution

The Industrial effluents are generally discharged to surrounding land mass or water bodies. These effluents contain several heavy metals that may prove harmful to water and soil environment. This causes land degradation and environmental pollution. The problem will further aggravated with increase in industrial activity. On long run, these toxic elements accumulate in soils and may cause health related problems to human beings and animals by entering into food chains.

1. Effluents from chromite mines

The state of Odisha has maximum deposits amounting 185 million tonnes of chromite ores spread over an area of 9197 ha at Sukinda in Jajpur District of

Odisha. This contributes to 98.4% of total chromite reserve of India. Mining and beneficiation of chromite ore at mining site cause a serious damage to the nearby vegetation, field crops, water bodies, human and animal health. Presence of high concentration of trivalent chromium (Cr^{3+}) in soil do not affect crops. But, when oxidized to hexavalent chromium (Cr^{6+}) become toxic to plants and humans. Presence of hexavalent chromium above 0.02 ppm in water is toxic. Toxic amount of hexavalent chromium causes nausea, skin diseases and lung cancer in human body.

A study carried out by the Department of Soil Science & Agricultural Chemistry, OUAT, Bhubaneswar showed that toxic amounts of hexavalent chromium was present in water bodies around the mining areas and its concentration decreases with increase in distance from the mining site (Table 10.11). Among the water bodies, rivulet, standing water in rice field and pond water contain higher concentration of Cr^{6+} as compared to tube well water or open well (Table 10.12).

The soils within 1-2 km radius from the mining site contain higher amount of Cr^{6+} and its concentration decreases with increasing the distance from the centre (Table 10.13).

Table 10.11. Hexavalent chromium content (Cr^{6+}) in water samples collected around mining area.

Distance (km)	Cr^{6+}(ppm)	
	Range	Mean
1	0.017-1.94	0.40
2	0.008-0.74	0.23
3	0.013-0.57	0.15
4	0.004-0.57	0.02
5	0.008-0.04	0.015
9	Trace - 0.08	0.025
15	0.03-0.035	0.23

Table 10.12. Hexavalent chromium (Cr^{6+}) content in different water bodies around the mining area.

Source	No. of samples	Cr^{6+}(ppm)	
		Range	Mean
Tube well	16	Trace – 0.068	0.021
Well	9	0.004-0.005	0.027
Pond	7	0.008-0.374	0.161
Rivulet	5	0.093-0.313	0.178
Flood water in rice field	5	0.025-0.465	0.189
Mine dam	3	0.550-0.741	0.619

Table 10.13. Hexavalent chromium content of soil around 0-15 km radius of mining site.

Distance(km)	Cr6+(ppm)	
	Range	Mean
1-2	0.399-10.07	0.61
3	0.216-0.962	0.44
4	Trace – 1.129	0.35
5	0.192-0.745	0.35
6	Trace – 0.433	0.14
8	0.120-0.553	0.31
9	0.048-0.288	0.13
11	Trace – 0.192	0.11
15	0.024-0.192	0.11

2. Effluents from tannery and textile industry

Tannery industry

India produces 100 million pieces of skins and hides which forms 12% of output of whole world. There are a total of 2161 ternaries in India which process 5 lakh tonnes of hides and skin annually. Maximum effluents produced is about 300 lit per 100 kg hide. Total annual discharge of effluents is about 94.2 lakh m^3

The leather processing requires large quantities of chemicals like sodium chloride, chromium sulphate, calcium and ammonium salts and organic dyes. The tannery effluents are ranked as the highest pollutants among all industrial effluents. The effluents from tannery industry generally contains much higher suspended solids, dissolved salts, chloride and chromium.

The accumulation of tannery effluents has undesirable effects on soil properties. Presence of excess amounts of sodium chloride causes deflocculation of soil particles. Sulphuric acid and salts contributes COD of tannery effluents. Trivalent chromium in effluents when enters into environment may not be toxic unless it is oxidized to hexavalent chromium. The tolerance limits for disposal of tannery effluents are given table 10.14.

Table 10.14. Tolerance limits for disposal of tannery effluents.

Parameters	Effluents to be discharged	
	Into surface water	On land for irrigation
pH	6-9	6-9
BOD (mg L^{-1})	30	100
Suspended solids (mg L^{-1})	100	200
Chloride (Cl) (mg L^{-1})	1000	200
Chromium (as Cr) (mg L^{-1})	2.0	2.0
Sulphides (as S) (mg L^{-1})	2.0	-
Sodium (%)	-	60
Oil and grease (mg L^{-1})	10	10

Source: Environmental Protection Rules (1986) In Pollution control Acts, Rules and Notifications Vol. 1. Central Pollution Control Board, New Delhi.

Textile Industry

In the textile industry, a number of chemicals including sodium salts, bleaching powder and detergents are used. The effluents produced from textile industry have high salinity, BOD, COD and excessive concentrations of sodium and carbonate ions. Large quantities of these effluents are discharged into the nearby rivers which cause serious geo-environmental problems. The textile effluents even on short term basis increases the amounts of water soluble salts, organic matter, Na and Ca resulting high SAR and RSC. Salinization and alkalization of ground water has been reported due to textile hand processing industries.

3. Paper mill and distillery effluents

In India, large amounts of effluents generated from distillery and paper mills are discharged on land or into the running water. The effluents are characterized by high suspended solids, BOD, COD and lignin content. The range of pollutants generated per tonne of the product is estimated at BOD 54-179 kg, suspended solids 109-280 kg and COD 160-488 kg. The standards for discharge of effluents is given in table 10.15.

Table 10.15. Standard of discharge of paper mill effluents.

Parameters	Value
pH	5.5-9.0
Suspended solids (mg L^{-1})	100
BOD (mg L^{-1})	100
COD (mg L^{-1})	250
Sulphide (mg L^{-1})	2.0
Residual chlorine (mg L^{-1})	1.0

The use of paper and distillery effluents indicates a significant increase in EC, organic carbon, Exch. Na as well as available N, P and K in soils.

The paper mill effluents adversely affects the seed germination and seedling growth of various crops. Disposal of effluents into water bodies increases organic load, depletion of oxygen and destruction of aquatic life. However, its application to sugarcane crop increases pH, O.C. and N, P and K status of soil. It was observed that the application of nutrients in form of effluents from distillery plants may be more effective to plants as compared to in organic fertilizers.

Mechanism of interaction of wastes with soil

The interaction mechanism of wastes in soil can be grouped under:

i) Physical – Filtration, dilution

ii) Chemical – Adsorption, precipitation, in organic transformations

iii) Biological – Organic matter decomposition, assimilation.

The infiltration of waste effluents in soil is influenced by the pore size distribution, particle size and organic matter content. The retention and transmission of dissolved ions and compounds are governed by the cation exchange capacity of soils. Organic particles, iron and aluminium particulate materials clogged the soil pores resulting surface accumulation and precipitation of heavy metals in rhizosphere zone.Specific adsorption occurs when soils retain certain cations and anions so tightly that they are not exchangeable.

It is reported that heavy metals are associated with soil solids in four major ways

i) A small portion is held in soluble or exchange forms, which are readily available to the plants,

ii) The elements like Cu and Cr are bound by the soil organic matter in the sludge which are not readily available to plants, but released over a period of time. But, Pb is not so highly attached.

iii) Association of heavy metals with carbonates and oxides of Fe and Mn are less available to plants.

iv) The residual form consists of sulfides and other insoluble compounds and are less available to plants than any other forms.

The accumulated heavy metals such as Zn, Cu, Ni and Cd in the root zone had severe adverse effect on plant growth. Update and accumulation of these heavy metals in toxic levels pose serious health hazard to humans and animals.

The biological transformation of wastes in the root zones adversely affect microbial activity. The biological diversity in waste water enhance the soil's ability to degrade a wide variety of substances. The amount of oxygen present in soil largely determines the amount of organic matter in soil. Initially the organisms digest more easily decomposable materials. This leads to formation of humus.

The Nitrogen content and soil permeability limit the waste application rates in soil. Domestic and municipal wastes, distillery effluents rich in nitrogen contributes imbalance C:N ratio in soils. The safe use of wastes in soils depends upon its BOD, suspended solids, N, P and quantity of toxic materials like Zn, Cu, Ni and Cd present in waste.

Pesticide Residues

The pesticides (insecticides, fungicides, herbicides) constitute an important input in crop production. In India the pesticide consumption is increased at the rate of 2 to 5% per annum. The consumption of pesticides is low in India (0.5 kg ha^{-1}) as compared to 1.0, 6.6 and 12.0 kg ha^{-1} in USA, Korea and Japan, respectively. Still there is a wide spread contamination of soils, waters and foods with pesticides. It was reported that 51 to 60 % of food items were contaminated with pesticides and 11 to 20% had residues above the maximum residue limit.

The higher pesticide contamination is due to indiscriminate use of chemicals in certain crops. About 84% of the pesticides were used in cotton, rice and vegetable and fruits covering one third of the cropped area (Table 10.16).

Table 10.16. Use of pesticides by different crops in India.

Crop	Pesticide used (%)	Cropped area %
Cotton	54	5
Rice	17	24
Vegetables and fruits	13	3
Plantation crops	8	2
Sugarcane	3	2
Oil seeds	2	10
Others	3	54
Total	**100**	**100**

Source: Agnihotri (1999)

The daily intake of contaminated food could be hazardous causing carcinogenicity, reduce life span and fertility, increased cholesterol, high infant mortality and many metabolic and genetic disorders.

Based on their chemical nature, the pesticides are classified as organophosphates, organochlorine compounds and carbonates having different mode of action stomach poison, contact poison, fumigants etc.

A large number of pesticide compounds have high immobility in soils and therefore, build up to toxic levels with their repeated use, especially in soils with low organic matter content and less microbial bio-diversity. Some of the herbicides like atrazine may persists in soils for 300-700 days under temperate climates.

Pesticide degradation in surface soil is affected by soil type, pH, moisture, temperature, organic carbon content, aeration, CEC and microbial diversity. Increased persistence in surface soil increases the possibility of ground water contamination. Pesticide sorption, degradation and water movement are the three process which control pesticide transport to the ground water. In aquifers, less of bacterial population, low nutrients and oxygen levels are the limiting factors for biodegradation of pesticides. The biochemical degradation by soil microorganisms is the single most important mechanism that can remove insecticide residues from the soil.

The integrated pest management (IPM), based on the synergistic use of resistant varieties, appropriate cultural practices, application of bio-pesticides and use of bio-control agents is an alternative eco-friendly approach to minimize pesticide contamination. The impact of IPM technology is quite evident in decreased consumption of pesticides in India.

Nitrate pollution

Soils heavily fertilized with N fertilizers above the recommended dose and infertile coarse textured soils are potential source of nitrate contamination in ground or run off waters. The amount, source, time and method of application of nitrogen fertilizers govern the effects of nitrate concentration in plants and ground water. Nitrate is absorbed in large amounts by plants such as fodder, vegetables, and result in nitrate toxicity to the consumers.

The leaching of nitrate is controlled by soil texture, rainfall amount and intensity, organic matter content in soil. Maximum leaching of nitrate below the root zone occurs in coarse textured soils of humid and semi humid regions. The World Health Organisation's standard for drinking water is 10mg NO_3-N L^{-1}.

The traditional animal waste disposal practices and the presence of manure pits on the village areas possibly create potential sources of nitrate pollution of ground water. There are several other sources of NO_3 –N namely sewage effluents, animal excreta, natural soil nitrates and decomposition products of soil organic matter. According to various surveys in India and Africa about 20-50% of well water contains nitrate levels greater than 50mg L^{-1} and, in some areas, as high as several hundred mg L^{-1}.

Excess nitrate taken in with water or food can result in methemoglobinemia, commonly known as the blue-baby syndrome in infants. Nitrate combined with haemoglobin forms methemoglobin that prevent oxygen transport. About 50 % infant death occurs due to oxygen starvation. The effects of nitrate on livestock are similar to human beings. In ruminants, rumen micro-organisms can easily reduce nitrate to nitrite which leads to methemoglobinemia. Cattles are more susceptible to nitrate poisoning. An excessive uptake of foods materials rich in nitrate (leafy vegetables) may increase the risks of cancer in human.

Reducing heavy metal pollution in soil

Heavy metal contamination in soils can be reduced either by reducing the application of toxins to the soil or by managing the soil- crop systems in a manner that leads to prevention of further recycling of the toxin. The cycle of toxic elements in soil - plant-animals is broken by immobilizing the toxins in soil. The soil becomes a sink for the toxic elements.

i) Liming the soil to neural pH level will reduce the availability and mobility of toxic elements. Liming reduced the uptake of Cd, Ni and Zn in plants but did not change the Cu content. The effect of liming in decreasing Cd uptake has been attributed to both in mobility of Cd in soils and competition between Ca^{2+} and Cd^{2+} on the root surface of absorption.

ii) Heavy phosphate application to soil can also lead to reduced availability of toxic cations.

iii) Biotechnology holds as promising role in pesticide waste disposal. The majority of pesticides can be degraded by fungi and bacteria. *Flavobacterium spp.* can rapidly hydrolyse organophosphate insecticides to less toxic level. *Trichoderma viride* can *hydrolyse malathion. Aspergillus niger* can degrade herbicides.

iv) Biosorption refers to adsorption of heavy metals to living or dead cells. Bacteria, yeasts, algae, fungi are very efficient in biosorption due to charged groups present in them. *Pseudomonas fluorescence* immobilized on polyvinyl chloride granules in a packed column was used to remove NO_3, Zn, Cr, Al, Ni, Pb from effluents.

Accumulation of heavy metals by cynobacteria removes Cd, Pb, Cu. Algal biomass is capable of binding considerable amounts of heavy metals like Zn, Cu, Cd, Pb. *Aspergillus oryzae* is used for removal of Cd with 99 % efficiency.

Some of the hyper accumulators used for metal remediation are given in table 10.17.

Table 10.17. Important and widely reported hyper – accumulators used for metal remediation.

Elements	Plant species	Max. conc. (mg kg^{-1})
Copper	*Ipomea alpina*	12300
Lead	*Brassica juncea and Zea mays*	8200
Nickel	*Alyssum lesbiacum*	47500
Zinc	*Brassica juncea, B. oleracea, B.campestris*	51600
Selenium	*Brassica junceaB. napus*	900
Chromium	*Brassica juncea*	1400

References

Agnihotri, N.P. (1999) Pesticide safety evaluation and monitoring. AICRP on Pesticide Residues. IARI, New Delhi, P. 173

ISI (2002) Draft standards for organic fertilizers (compost).

Jena, D; Nayak, M.K. Acharya, N. and Singh, M.V. (2003) Fluoride distribution in soil, water and plant in the vicinity of NALCO smelter plant at Angul, Odisha. In. Environmental Pollution. Edited by Vijay P. Singh and P.N. Yadav. Allied Publishers Pvt. Ltd. PP: 188-194.

Juwarkar, A.S; Juwarker, Asha; Deshbhratar, P.B. and Bal, A.S. (1991) Exploitation of nutrient potential of sewage sludge through land application. In. Asian Experience in Integrated Plant Nutrition. RAPA, FAO of the United Nations, Bangkok, PP. 178 – 201.

Sanyal, S.K. (2001) Final Report. ICAR Adhoc project, "Status, casues and impacts of Arsenic contamination in ground water in parts of West Bengal Vis-à-vis management of agriculture system."

Sengupta, S.R. and Pal, B. (1971) Iodine and fluoride content of food stuffs. Indian J. of Nutrition and Diabeties. 8: 66-71.

Study Questions

1. How do the modern agricultural activities pollute the soil?
2. What are the sources of different heavy metals in the soil?
3. Discuss different point and non-point sources of water pollution.
4. What is the difference between pollution and contamination.
5. Write short notes on
   i) Acid rain
   ii) Eutrophication
   iii) Biosorption

iv) Nitrate pollution

v) Biological magnification

vi) Photochemical smog

6. Name the contaminants causing following diseases

   i) Minamata disease

   ii) Itai – Itai

   iii) Blue – baby syndrome

   iv) Black – foot disease

   v) Selenosis

   vi) Fluorosis

7. Name the primary and secondary air pollutants and their sources.

8. Briefly discuss the effect of arsenic toxicity on soil and crops. Give some remedial measures.

9. Which form of selenium is more toxic.What are the toxic effects of selenium on animals and human.

10. Name different fluoride containing minerals. How these minerals contaminate ground water?

11. Briefly discuss the extent of fluoride toxicity in water, soil and plants around an alluminium smelter plant.

12. State the beneficial and harmful effect of urban sewage water on crops.

13. Explain the mechanism of interaction of wastes with soil.

14. Discuss the impact of tannery industry and paper mill effluents on soil properties.

11

Soil Quality and Soil Health

Soil health

Soil health is defined as the continued capacity of soil to function as a vital living system, by recognizing that it contains biological elements that are keys to ecosystem function within land use-boundaries. These functions are able to sustain biological activity of soil, maintain the quality of air and water as well as promote plant, animal and human health.

The characteristics of a healthy soil depends on the physical, chemical and biological properties of soils which directly or indirectly influence crop production.

Soil quality and soil health are terms that are discussed in agricultural community. Both these terms encompass a complex, dynamic concept, different to define and hard to measure. The term soil quality and soil health are often used interchangeably in the scientific literature, with scientists in general preferring soil quality and producers preferring soil health. Soil health may be considered as the state of soil at a particular time equivalent to the dynamic properties in short term. In contrast, soil quality may be considered for a particular purpose over a longer time scale, equivalent to static soil quality. In a broader way soil quality may refer to both permanent and static soil properties. (Fig. 11.1).

A static soil property relates to natural characteristic of soil (texture) which are derived from pedogenic processes. These properties cannot be easily amended. The dynamic properties are readily related to management practices. The dynamic component is of most important for the growers because it allows the soil to come to its full potential.

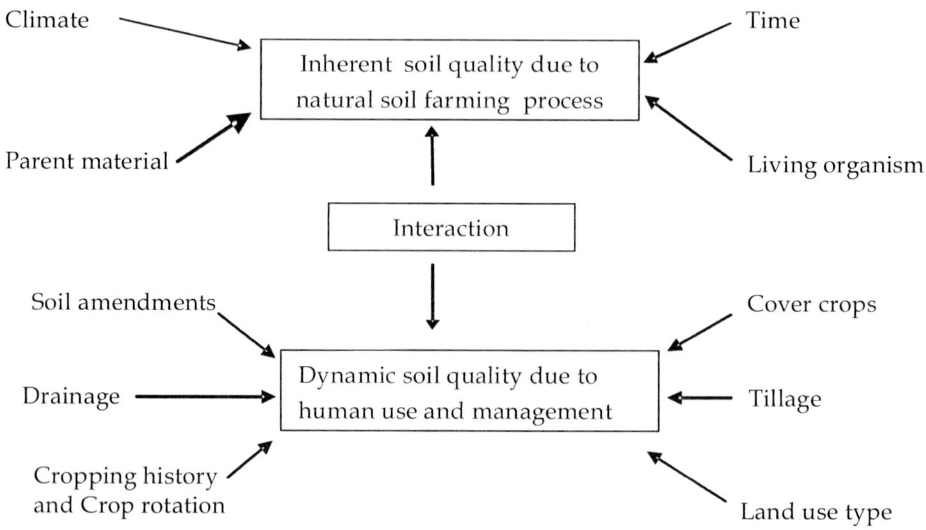

Fig. 11.1. Inherent and dynamic soil quality and factors affecting them

Soil health is the integration of soil physical, chemical and biological process and functions. A healthy soil will be balanced for all three components (Fig. 11.2).

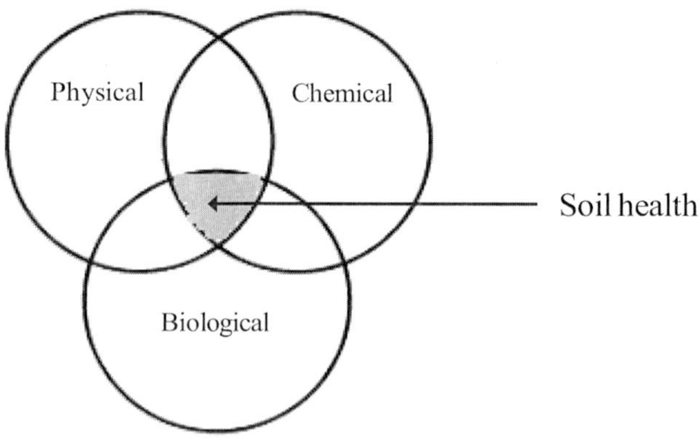

Fig. 11.2. The concept of soil health

Soil quality

The term and concept of soil quality evokes various responses depending on our scientific and social backgrounds. Historically, soil quality meant suitability or limitations of a soil for a particular use. The concept of soil quality has been broadened to include attributes of food safety and quality, human and animal health and environmental quality.

Karlen et al. (1977) defined soil quality as "The capacity of a specific kind of soil to function, within natural or managed ecosystem boundaries to sustain plant and animal productivity, maintain or enhance water and air quality, and support human health and habitation."

Doran et al. (1996) defined soil quality as "The capacity of living soil to function within natural or managed ecosystem boundaries, to sustain plant and animal productivity, and promote plant and animal health."

Presently, soil quality has been defined by scientists as the, "fitness for use"

Soil function

Karlen et al. (1977) described five soil functions as:

i) Regulating water

ii) Sustaining plant and animal life

iii) Filtering potential pollutants

iv) Cycling nutrients

v) Supporting structures

Collectively, the soil functions are evaluated with respect to their capacity to sustain plant and animal productivity, maintain to enhance water and air quality, and support human health and habitation in determining soil quality.

Evaluation of soil quality

Evaluation of land needs to be dynamic and to reflect the response of the natural environment to change in input and production objectives. FAO (1977), outlines the principles for land appraisal as:

- Land suitability is only meaningful in relation to a specific use
- Suitability must refer to use on sustained basis
- Evaluation involves comparison of more than one alternative type of land use.
- Different kinds of land use are compared at least on a single economic basis.

***Eswaran and Virmani* (1990) listed following as being an integral part of sustainability of agriculture:**

- Buffered against risks
- Established over time
- Synchronized with climate

- Harmonized with environment
- Sensitive to soil degradation
- Response to change

They further reported that soil qualities important for sustainable agriculture are : available water capacity, nutrient retention capacity-CEC, nutrient availability, pH and base saturation, chemical constraints-acidity, sodicity, salinity, and physical constrains-low hydraulic conductivity, permeability, high bulk density, crusting, surface tilth, erodibility, water logging. These authors further suggested, it is necessary to establish threshold values with respect to specific uses for each of the soil qualities for sustainable agriculture.

Lal (1994) proposed several indices of measuring sustainability:

- **Productivity P = p/R**where, P – Production per unit resources

  p – Grain yield (Say)

  R– Resource in put
- **Total Factor Productivity (TFP)**

  TFP = P/ S(Ri :Ci) where,

  Ri – Resource used

  Ci – Cost of specific input
- **Total Natural Resource Productivity (TNRP)**

  TNRP =TFP /D S where,

  TFP – Total factor productivity

  D S– Change in critical soil property, which may be annual soil loss, SOM, nitrate concentration etc.
- **Storie Index Rating (SIR):** SIR is an old land evaluation system developed in US (Storie 1993, 1976). It was developed for the agricultural rating of land for tax assessment and land use zoning. It does not include management factors. The SIR is assessed by assigning percentage rating to each factor (ideal soil condition is 100%) and then multiplying these rating as:

  Storie Index = A x B x C x D where,

A – Soil factor (depth, permeability, and gravel content of the soil)

B – Texture factor

C – Slope factor

D – Miscellaneous factor

The index is normalized to a percent rating.

Soil Quality Indicators

Soils have physical, chemical and biological properties that interact in a complex way to allow the soil to do its function. Soil quality cannot be measured directly, but must be inferred from measuring changes in its attributes of ecosystem, referred to as indicators. Indicators of soil quality should give some measure of the capacity of the soil to function with respect to plant and biological productivity, environmental quality and human and animal health. Indicators are measurable properties of soil or plants that provide clues about how well the soil can function.

Essential Characteristics of Soil Quality Indicators

- Should be easy to measures
- Should measure changes in soil function
- Encompass physical, chemical and biological properties
- Should be accessible to many users and applicable to field conditions
- Should be sensitive to variations in climate and management
- Should be cost effective
- Should have low error with measurement
- Should have community acceptance and involvement
- Indicators can be assessed by qualitative or quantitative techniques.

Importance of Indicators

Soil quality indicators are important to

- Focus conservation efforts on maintaining and improving the condition of soil
- Evaluate soil management practices and techniques
- Relate soil quality to that of other measures
- Collect the necessary information to determine trends
- Determine trends in the health of the Nation's soil
- Guide land manager decisions

Criteria for Selecting Indicators

The selection of indicators should be based on:

- The land use
- The relation between an indicator and the soil function being assessed

- The ease and reliability of the measurement
- Variation between sampling time and variation across the sampling area
- The sensitivity of the measurement to changes in soil management
- Compatibility with routine sampling and monitoring
- The skill required for use and interpretation

Important Soil Quality Indicators

Soil quality cannot be measured directly. Therefore, we have to evaluate indicators which are measurable soil properties that provide clues about how well the soil function. The indicators that directly monitor the soil are grouped into four categories as visual, chemical, physical, and biological indicators.

i) Visual Indicators of Soil Quality

Visual indicators of soil health may be obtained from observation or photographic interpretation. Exposure of sub soil, change in soil color, water logging, run off, plant response, weed species are a few examples of visual indicators. Visual evidence can give a clear indication of soil quality.

ii) Physical Indicators of Soil Quality

Physical indicators of soil health reflect the capacity to accept, store, transmit and supply water, oxygen and nutrients within ecosystem. This includes monitoring of soil structure through pore size distribution, aggregate stability, saturated hydraulic conductivity, infiltration, bulk density and surface crust. Rooting depth provides a good indicator of buffering against water, air and nutrient stress. Soil surface cover can be used as an indicator of surface protection against rain drop impact, splash erosion, reduce surface crust and hence enhanced infiltration (Table 11.1).

Table 11.1. Predominant soil physical attributes and related soil process.

Attributes	Processes
Texture	Crusting, gaseous diffusion, infiltration
Bulk density	Compaction, root growth, infiltration
Aggregation	Erosion, crusting, infiltration, gaseous diffusion
Poresize distribution	Water retention, and transmission, root growth, gas exchange
Available water capacity	Drought stress, biomass production, soil organic matter content
Non-limiting water range	Soil structure, drought, water imbalance
Infiltration rate	Run off, leaching, erosion
Effective rooting depth	Root growth, nutrient and water use efficiency
Soil temperature	Heat flux, microorganism population and activity

Source: Lal (1994)

Infiltration

Infiltration refers to the entry of water into soil at soil-atmosphere interface. Deterioration of soil structure, presence of exchangeable sodium (sodicity), loss of organic matter decreases the rate of infiltration. Poor infiltration rate causes water stagnation and run off whereas high rate of infiltration leads to nutrient loss.

Aggregate Stability

Aggregate stability refers to the resistance of soil aggregates to break down by water and mechanical force. Aggregate stability is influenced by quantity of organic matter, types of clays, amount of electrotype, biological activity, green manuring, cropping pattern and tillage practices.

Bulk Density

Bulk density of soil is defined as the weight of oven dry soil with respect to volume of soil including pore space. It is a dynamic property of soil which is easily changed with management practices. It gives useful information about porosity of soil, total water storage capacity, root penetration and proliferation. The bulk density values are rated as non-limiting, critical and root limiting (Table 11.2).

Rooting Depth

Rooting depth is a good indicator of plant available water capacity. Sub soil salinity, soil compaction affects rooting depth.

Table 11.2. Bulk density (Mgm^{-3}) limits for different textural classes.

Textural class	Non limiting bulk density	Critical bulk density	Root limiting bulk density
Sandy	1.60	1.69	1.85
Coarse-loamy	1.50	1.63	1.80
Fine-loamy	1.46	1.67	1.78
Coarse –silty	1.43	1.67	1.79
Fine – silty	1.34	1.54	1.65
Clayey	1.40	1.49	1.58

Water Holding Capacity

Water holding capacity of soil is the amount of water a soil can hold after rainfall or irrigation. The water holding capacity of soil is influenced by soil structure, texture, percent of clay, organic matter content and soil aggregation.

Fine textured soils hold more water than coarse-textured soil. Organic soil can hold more water as compared to mineral rich soils. The major management practices that influence water holding capacity are tillage, residue management.

iii) Chemical indicators of soil quality

Chemical indicators can give information about the equilibrium between soil solution and exchange sites, nutritional requirements of plant, levels of soil contamination with heavy metals (Table 11.3).

Chemical Indicators Include

- Organic matter measures such as total organic carbon and total nitrogen
- Soil fertility
- Soil reaction (pH)
- Measures of salinity or sodicity such as electrical conductivity, SAR and ESP
- Measures of toxic contaminants

Organic matter content of soils is an essential component of soil fertility and productivity in many ways such as

- It provides the carbon and energy sources for microbes and earth worms
- Improves soil structure
- Store and transmit water and air
- It stores and supplies plant nutrients such as N, P, S and micronutrients
- It enhances CEC of soils and retain nutrients
- It makes the soil more friable, less sticky and reduces compaction
- It retains carbon in stable form
- It reduces the contamination of pesticides, pollutants

Loss of organic matter from soil due to continuous cropping results in deterioration of soil quality and productivity.

Soil pH is a measure of the soil acidity and alkalinity. Soil pH is directly related to plant growth as it affects plant nutrient availability and microbial activity. A pH range of 6 to 7 is ideal for plant growth because most of the plant nutrients are readily available in this pH range.

Electrical conductivity (EC) provides an assessment of salinity status of soil. Electrical conductivity of less than 4 dS m^{-1} is ideal for plant growth. Higher the

EC, greater is the salinity hazard.

Besides the above chemical indicator of soil quality, CEC, ESP, concentration of heavy metals and pesticides has been suggested for specific situations.

iv) Biological indicators of soil quality

Biological indicators of soil health includes soil microbial biomass carbon and respiration, potentially mineralizable N, enzyme activity, microbial diversity, nematode and earthworm populations (Table 11.3).

Soil microbial biomass carbon

Soil microbial biomass carbon is a labile source and sink of plant nutrients. It regulates nutrient availability and nutrient cycling and is a good indicators of potential microbial activity,.

Besides, microbial biomass carbon, it is also suggested to monitor microbial quotient and microbial metabolic quotient. Microbial quotient is the ratio of microbial biomass carbon to total soil organic carbon. Microbial metabolic quotient or respiratory quotient is the ratio of rate of respiration to microbial biomass carbon.

Potentially mineralizable N

Potentially mineralizable N is measured by either aerobic or anaerobic incubation. Potentially mineralizable N measures N supplying capacity and also indicates a labile fraction of soil organic matter.

Enzyme activity

Soil enzyme activity is often closely related to soil organic matter, microbial activity and microbial biomass. Potentially useful indicators of soil quality could be beta-glucosidase, urease, amidase and phosphates.

Earthworm and nematodes

In soil, nematodes and earth worm communities affect soil structure and influence organic matter dynamics and nutrient cycling.

Table 11.3. Predominant soil chemical and biological attributes and related process.

pH	Soil acidity and alkalinity, nutrient availability
Base saturation	Absorption, desorption, solubilization
CEC	Ion exchange, leaching
Total and available nutrients	Soil fertility, nutrient reserves
Soil organic matter	Soil structure formation, mineralization, biomass carbon, nutrient retention
Earthworm and other soil micro fauna	Nutrient cycling, organic matter decomposition, soil structure formation
Soil biomass carbon	Microbial transpiration and respiration, soil structure formation and organo mineral complexes
Total soil organic carbon	Soil nutrient sources and sink, biomass carbon, soil respiration and gaseous fluxes

Source: Lal (1994)

Development of soil quality indices

Soil Management Assessment Framework Design (SMAF) followed three steps: Indicator selection, Indicator interpretation, and integration into a soil quality index value (Fig. 11.3)

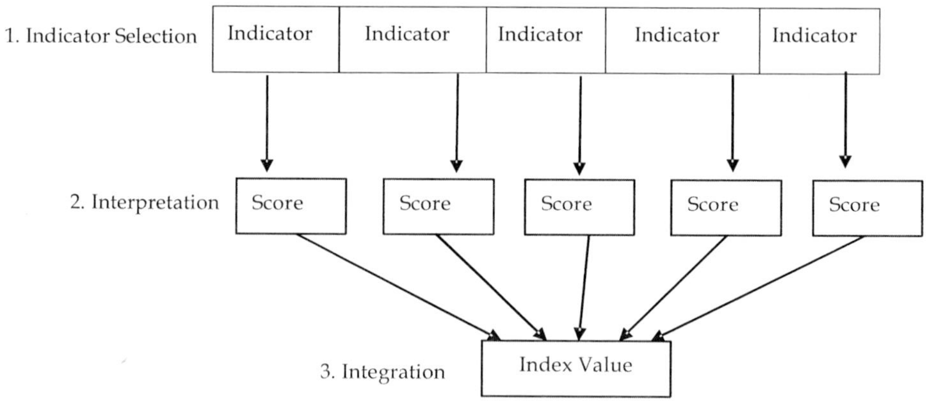

Fig. 11.3. Conceptual framework for the soil management tool.

Minimum Data Set (MDS)

A minimum data set is obtained through selection of key indicators such as texture, organic matter, pH, EC, bulk density, rooting depth, nutrient status. It identifies locally relevant indicators that influences the soil quality. Selection of minimum data set will help the farmers and extension workers to understand the problem and target the solution at farm level.

Lal. (1994) proposed a list of minimum data set for assessment of soil quality under different climatic regions (Table 11.4).

Table 11.4. Minimum data set for soil quality assessment under different regions.

Humid	Semi humid / semi arid	Arid
pH, base saturation	pH, soil organic matter content, available nutrients, salinity, soil structure, bulk density, compaction, soil tilth	Plant available nutrients
Bulk density, soil tilth, infiltration rate, available water capacity	Erosion by water and wind, soil air & temperature, growing season	Compaction, crusting, rooting depth, water availability, surface and ground water quality, soil and air temperature, evaporation and wind velocity.
Run off rate, water erosion Rainfall amount and intensity, net radiation, soil temperature		

Methods of calculating soils quality index

There are several methods of assessment of soil quality. The key indicator approach method is discussed here.

Key indicator approach

This method has been followed by Andrew and Carroll (2001).

The various steps involved in the approach have been discussed below:

Comprehensive analysis of various attributes

The first step in the process is the analysis of various physical, chemical and biological attributes, yield and water quality parameters.

Defines goals

The second step in evaluating soil quality is the identification of management goals or soil function under any eco system-management practice. The different data set for defining goals may be yield per unit land, return per unit land, organicmatter status under a specific management practice or sustainability yield index.

Choosing the minimum data set (MDS)

The determined soil quality indicators are reduced to a minimum data set through a series of uni and multivariate statistical methods using SPSS 10 software. Both parametric (RBD) as well as non-parametric statistics (Kruskal-Wallis X^2) are to be used to identify quality indicators with significant treatment differences. For each statistically significant variable Principal Component Analysis (PCA) may be performed. Within each principal component, only highly weighted factors those with absolute values within 10% of the highest weight are taken for the MDS. Reduction of redundancy and exclusion of spurious groupings of the indicators are made using inter correlation techniques between the variables. The multi variate correlation co-efficients may be used to determine the strength of relationship among variables. Variables should be eliminated on the basis of their relative importance towards functional goals using multiple regressions. Finally, the screened variables exhibiting significant contribution towards functional goals should be leveled as key indicators of soil quality. Besides this, any highly weighted variables can also be considered important and retained in the MDS.

Assigning the score values to the key indicators

After selection, the key indicators should be transformed using a scoring method for inclusion in soil quality index (SQI). Each of the variables should be scored based on their performance of soil function.

Every observation of the variables should be transformed for linear or non – linear scoring functions where the Y-axis ranges from 0 to 1 and the X- axis represents a site dependent expected range. A score value of 1 indicates high function.

For linear scoring, the data on indicators should be arranged in ascending or descending order depending on whether a higher value considers 'good' or 'bad' with respect to soil function. For 'more is better' indicators, each observation should be divided by the highest observed value such that the highest observed value receives a score 1. For 'less is better' indicators, the lowest observed value should be divided by each observation such that the lowest observed value receives a score of 1.

Computation of soil quality index (SQI)

Two soil quality indices such as an additive SQI and a weighted additive SQI may be used for a comparison.

The additive index is the summation of scores from MDS indicators. From the summed scores, the additive soil quality index treatment means standard deviations are to be calculated.

In the weighted additive index, after transformation, the MDS variables for each observation should be weighted using the PCA results. Each PC component would explain a certain amount (%) of variation in the total data set.

This percentage, divided by the total percentage of variation explained by all PCs with eigenvectors > 1, will provide the weighted factor for variables chosen under a given PC.

The weighted MDS variables score should be summed for each observation using the equation;

$SQI = \sum_{i=1}^{n} (wi \times Si)$ where,

Si is the score for the variable i and Wi is the weighing factor derived from PCA for the variable i.

In this process higher index scores indicate better soil quality or greater performance of soil function. Further, the percent contribution of each final key indicator should also be calculated. The SQI values so obtained should be tested for their level of significance.

A case study

The results of a long term tillage-residue-nitrogen management experiment in Alfisoil of Central Institute of Dryland Agriculture, Hyderabad is presented here for benefit of the readers regarding computation of soil quality index(Sharma et al. 2004). The experiment was laid out in split-split design with sorghum and castor rotation. The strip constituted two tillage treatments- minimum tillage (MT) and conventional tillage (CT). The main plot consist of three residue treatments – sorghum stover (SS), gliricidialoppings (GL) and no residence (NR) and sub plot consists of four level of N as N_0, N_{30}, N_{60} and N_{90}.

Average yield of sorghum and castor and sustainable yield index (SYI) over 8 years were defined as goals of this study.

The average yield of sorghum and castor was highest in CT GLN_{90} treatment. The SYI was significantly higher under CT than under MT. SYI was highest in CT SS T_{90}.The interactive effects of tillage x nitrogen and residue x nitrogen on SYI were significant.

Selection of Indicators

Soil responses to different management practices differently. Out of 21 variables (soil properties), tillage significantly influences 9 variables, residues influences 13 variables and N levels influences 17 variables. The key indicators for PCA analysis were chosen based on the criteria that a variable should have two or more significant effects on the treatments (Table 11.5).

Based on this criteria six soil properties viz. pH, Ca, Mg, Zn, Fe and available water (having one significant value) were dropped and remaining 15 parameters (with two or more significant value) were selected for PCA analysis.

Principal Component Analysis (PCA)

After determining the MDS indicators, every observations of each MDS indicator was transformed using a linear scoring method. Once transformed, the MDS variables for each observation were weighted using the PCA results. Each PC explained a certain amount (%) of the variation in the data set. This percentage divided by the total percentage of variation explained by all PCs with eigenvectors more than 1, provided the weighted factor for variables under a given PC.

In the PCA of 15 variables, five PCs had eigen value more than 1 and explained 78 % of the variance in the data. A correlation matrix for the highly weighted variables under different PCs was run separately. It was assumed that the variables having highest correlation sum best represented the group (Table 11.6). Among the four variables in PC_1, MBC and available N were retained in MDS because of highest correlation sum.

In PC_2, although K and HC were not correlated, but retained in MDS because of their relative importance in dry land agriculture.

Under PC_3 and PC_4, all the variables were dropped due to low variation between treatments.

In PC_5 available S was retained in MDS because of frequent deficiency of S in dry land Alfisol.

Hence, the final MDS consisted of five variables such as MBC, N, K, HC& S.

All the key indicators, retained in MDS were regressed as independent variables with data on management goals as dependent variables. The co-efficient of determination (R^2) with SYI, average yield of sorghum and castor were 0.411, 0.42 and 0.507, respectively (Table 11.7).

Multiple regressions revealed that N and HC significantly influenced all three management goals, while effect of S was only significant on SYI. Available K and MBC had significant influence on average castor yield.

Soil quality index

The soil quality index was calculated using the equation,

$$SQI=\sum_{i=1}^{n}(wi \times Si) \text{ where,}$$

S_i is the score for subscripted variable and Wi is the weighing factor derived from the PCA.

The SQI data presented in table 11.8 showed that main effects of residue and nitrogen levels on SQI were significant. Main interaction effects were significant. Application of glyricidialoppings was found superior to sorghum stover and no residue in maintaining higher SQI values. Increasing N levels also maintain higher SQI values. Among the combination of 24 treatments, the SQI ranged from 0.90 to 1.27. highest being in CTCL N_{90} and lowest in MTNR N_{30}.

The contribution of individual indicators towards SQI was 32% for N, 31% for MBC, 17% for K and 4% for S.

Available N plays a dominant role in maintaining the yield of dry land crops. It helps in increasing the vegetative growth as well as root biomass. This in term enhances the organic matter content in soil. Soil micro-organisms are the driving force behind soil organic matter transformations such as, mineralization, immobilization of organic constituents. These transformations are the basis of plant decomposition, soil aggregation, soil tilth, and nutrient availability.

The contribution of HC to SQI is likely derived from the soil's tendency to crust formation & poor water retention. Available K is important for osmotic regulations.

Table 11.5. Effect of tillage, residues and nitrogen levels on soil quality indicators (correlation matrix).

S. No.	Soil quality Indicators	Tillage	Residue	Nitrogen	T x R	TxN	R xN	TxRxN
1.	pH	NS	NS	* *	NS	NS	NS	NS
2.	EC(dSm^{-1})	NS	* *	* *	* *	* *	* *	* *
3.	OC. (g kg^{1})	NS	* *	* *	* *	NS	* *	NS
4.	Av.N (kg ha^{1})	NS	* *	* *	NS	*	* *	* *
5.	Av.P(kg ha^{1})	NS	* *	* *	NS	* *	* *	* *
6.	Av.K (kg ha^{1})	* *	* *	* *	* *	* *	* *	* *
7.	Exch.Ca (cmol(+) kg^{1})	NS	NS	NS	NS	*	NS	NS
8.	Exch. Mg cmol(+)kg^{1})	*	NS	NS	NS	NS	NS	NS
9.	Av. S (kg ha^{1})	NS	*	**	**	NS	* *	* *
	Available (mg g^{-1})							
10.	Zn	NS	*	NS	NS	NS	NS	NS
11.	Fe	NS	NS	* *	NS	NS	NS	NS
12.	Cu	*	NS	* *	* *	* *	* *	* *
13.	Mn	*	NS	* *	NS	NS	NS	NS
14.	B	NS	*	* *	NS	NS	* *	* *
15.	MBC (mg g^{-1})	* *	* *	* *	* *	NS	* *	* *
16.	DHA (mg TPF 24h^{-1} g^{-1})	NS	* *	* *	* *	* *	* *	* *
17.	MBN (kg ha^{-1})	*	* *	* *	* *	* *	* *	* *
18.	BD (Mg kg $^{-1}$)	**	* *	* *	* *	* *	* *	* *
19.	HC (cm h^{-1})	*	NS	NS	*	NS	NS	* *
20.	MWD (mm)	**	* *	* *	* *	* *	* *	* *
21.	AW (w/w)	NS	NS	* *	NS	NS	NS	NS

* and ** indicates significant difference of 5 % and 1%, respectively and NS – non significant. MBC – Microbial biomass carbon; DHA – Dehydrogenase activity, MBN – Microbial biomass nitrogen, HC – Hydraulic conductivity, MWD – mean weight diameter, AW – available water.

Table 11.6. Correlation matrix (Pearson's co-efficient) for highly weighted variables under PC's with high factor loading.

PC variables	Organic carbon	N	MBC	MBN
PC 1 variables				
Organic Carbon	1.000	0.479*	0.846**	0.550**
N	0.479*	1.000	0.625**	0.829**
MBC	0.846**	0.625**	1.000	0.763**
MBN	0.550**	0.829**	0.763**	1.000
Correlation sums	2.875	2.933	3.234	3.142
PC 2 Variables	K	HC		
K	1.000	0.334		
HC	0.334	1.000		
PC3 variable	EC			
EC	1.000			
PC 4 variables	EC	Cu	DHA	
EC	1.00	-0.104	-0.203	
Cu	-0.014	1.000	0.319	
DHA	-0.203	0.319	1.000	
Correlation Sums	1.217	1.423	1.522	
PC 5 variable	S			
S	1.000			

* and ** indicate significant difference at 5% and 1%, respectively.

Table 11.7. Multiple regression of the MDS components using management goal attributes at different probability (P) levels.

Goal or function	R^{2*}	MDS variables	P
SYI	0.411	N, HC, S	>.000, >0.000, >0.035
Average sorghum yield	0.420	N,HC	>.000, >0.001
Average castor yield	0.507	N,K,MBC, HC	>0.000, > 0.016, >0.049, >0.000

* Significant at P = 0.01 level.

Table 11.8. Effect of different land management treatments on SQI.

N levels (kg ha^{-1})	CT			MT		
	S S	GL	NR	S S	GL	NR
N0	1.02	1.10	0.95	1.12	0.99	0.94
N30	1.07	1.12	0.97	1.11	1.11	0.90
N60	1.17	1.19	1.01	1.13	1.15	1.01
N90	1.16	1.27	1.02	1.18	1.17	1.10
Tillage (T)	NS					
Residue (R)	**					
Nitrogen (N)	**					
T x R	**					
T x N	NS					
R x N	**					
T x R x N	**					

NS: Non Significant, ** Significant at 1 %

Strategies for improving soil quality and sustainability

The strategies for maintaining and improving soil quality must focus on:

1. Reducing further degradation and
2. Counteracting the ill effects of degradation already occurred.

Adoption of management practices such as mulch farming techniques, conservation tillage, cover crops and green manuring diversified cropping system, nutrient cycling mechanism, periodic application of organic matter helps to maintain soil health and soil quality. Further, deterioration in soil quality can be effectively checked by following appropriate land use system, suitable crop rotation, soil test based fertilizer application, correction of soil acidity through liming and sodicity by gypsum application.

Maintenance of soil structure by application of organic matter, conservation tillage, mulch farming resulted in easy movement of air, water and gases.

Growing of deep rooted crops, sub-soiling, plowing at appropriate moisture content reduces soil compaction.

For enhancing soil fertility, balanced fertilization, growing of legumes and recycling of organic waste are recommended.

Integrated use of fertilizers alongwith farmyard manure, compost, green manure or crop residues enhanced the total organic carbon status of soil which has greater implication on carbon sequestration.

To keep track on soil quality, soil health cards should be issued to the farmers so that farmers can periodically observe the changes in some of the visible indicators of soil health.

References

Andrews, S.S. and Carroll, C.R. (2001) Designing a decision tool for sustainable agroecosystem management: Soil quality assessment of a poultry litter management case study.Ecol. Appl. 11(6).

Doran, J.M., Sarrantonio, M. and Liberg, M.A. (1996) Soil health and sustainability. Adv. Agron. 56:1-54.

Eswaran, Hari and Viramani, S.M. (1990)The soil component in sustainable agriculture. In Proceeding of the International Symposium on Natural Resource Management for a Sustainable Agriculture. Indian Society of Soil Science, New Delhi. pp. 64-86.

FAO (1977) A Frame Work for land evaluation. Food and Agriculture Organization. Soil Bulletin, 32, FAO, Rome.

Karlen, D.L., Mausbach, M.J., Doran, J.W; C line, R.G., Harries, R.F. and Schuman, G.E. (1997) Soil quality. A concepts, definition and framework for evaluation. Soil Science Society of America Journal 6:4-8.

Lal, R. (1994) Methods and Guidelines for Assessing Land use of Soil and Water Resources in the Tropics. SMSS Technical Monograph No. 21. Soil Management Support Services, Washington DC, USA.

Storie, R.E., (1976) Storie Index Soil rating (Revised 1978) special publication. Division of Agricultural Sciences. Number 3203, University of California, Berkeley, CA, USA.

Storie, R.E. (1993) An Index for rating the agricultural values of soils (Revised 1937). University of California Agriculture Experiment Station Bulletin 556. Berkeley, CA, USA.

Sharma, K.L., Srinivas, K., Mandal, U.K., Ramakrishna, Y.S., Vittal, K.P.R.,

Venkateswarlu, B., Ramesh, V. and Kusuma Grace, J. (2004) Strategies and Approaches to Improve and Assess soil quality. NATP, Central Research Institute for Dryland Agriculture, Hyderabad.

Study Questions

1. Define soil quality. How it is different from soil health?
2. Differentiate between inherent and dynamic soil quality.
3. Are all the estimated soil properties considered as soil quality indicators? State the essential characteristics of soil indicators.
4. What are the impacts of intensive cropping on soil quality?
5. Name different physical, chemical and biological soil quality indicators and their associated functions?
6. Suggest some management practices to improve soil quality on long term basis.
7. Write in briefs the steps of assessment of soil quality index using key indicator approach.

12

Remote Sensing, GIS and GPS in Management of Problem Soils

Sustainable management of soil resources is essential for food, nutritional and economic security of the country. The continued degradation of soil resources is considered an important factor in lowering the total and partial factor productivity of agriculture in the country. The process like soil erosion, nutrient loss, water logging, desertification, acidification, compaction, crusting, salinization and soil contamination affect food and environmental security in most parts of the country. The future agricultural growth at higher rate cannot be sustained with a deteriorating soil resource base. We need to monitor the quality of soil resources and devise appropriate management strategies for sustainable crop production without deteriorating the environment.

In India, several agencies like National Commission on Agriculture (NCA), Ministry of Agriculture (Soil and Water conservation Division), Society for Promotion of Waste lands Development (SPWD), NBSS&LUP, Nagpur have been constantly inventorizing and characterizing soil resource for drawing up land use and development plans at National and State levels. However, there is a great variance in estimates due to differences in approaches and criteria for assessment. For an example,the total degraded land as estimated by NCA is 148.09 Mha as against, 175.0 Mha by Ministry of Agriculture (Soil and water conservation Division), 129.58 Mha by SPWD and 187.7 Mha by NBSS&LUP, Nagpur.

In soil profile development, we focused on vertical variability in soils - the differences among soil horizons. In this chapter we focused on horizontal variability- how soils differ from place to place across the landscape. Spatial variations in soil properties occurs across distances having geographic meaning for land management-from a few meters to many kilometers. We need to monitor regularly the status of our soil resources employing modern technologies to refine our understanding about them and renew our commitment to their preservation. The applications of Remote Sensing (RS), Geographical Information Systems (GIS) and Global Positioning System (GPS) have greater opportunities to offer in this regard.

Geoinformatics

Geoinformatics is an inter disciplinary field with Remote Sensing, GIS, GPS as its core subjects. Geoinformatics is related to spatial data acquiring, measuring, analyzing, storing, managing, displaying and applying it in "soil resource monitoring and management.

Remote Sensing

Remote Sensing describes the gathering of information about an object from a distance (i.e. not coming in contact with it).

According to the physics of Remote Sensing, different objects return different amounts of energy in different wave lengths of the electromagnetic spectrum. Detection and measurements of these spectral signatures enables identification of surface objects from the air-borne and space-borne platforms. This technology, integrated with traditional techniques, is an efficient, speedy, cost effective and important tool in the management and exploration of natural resources.

Principles

A Remote Sensing system using electromagnetic radiation has four components: a source, interaction with earth surface and atmosphere and a sensor. The source of electromagnetic radiation may be Sun's reflected light or Earth's emitted heat, or man made microwave radar. The amount and characteristics of radiation emitted or reflected from the earth's surface depend upon the characteristics of the object on the earth's surface. The electromagnetic energy passing through the atmosphere may be

i) Transmitted through the substance,

ii) Absorbed by a substances,

iii) Emitted by a substances,

iv) Scattered in all directions and lost and ultimately reflected.

The sensor detects and records changes in electromagnetic radiation by magnitude, direction, wavelengths, polarization and phase. The resulting images and data are interpreted remotely to identify the characteristics of the matter that produced the changes in the recorded electromagnetic radiation.

Basic Processes of Remote Sensing

Basic processes of remote sensing are as follows:

- Energy source (Sun or transmitter)
- Transmission of energy from source to object

- Energy interaction with object surface
- Transmission of energy to sensor
- Scattering and absorption by atmosphere
- Data acquisition, recording, processing and interpretation

Types of Remote Sensing

Remote sensors are the instruments which detects various objects on the earth's surface by measuring electromagnetic energy either reflected or emitted by them. Remote sensors can be classified in different ways as:

1. On the basis of source of energy used

a) Active RS

It supplies its own source of energy which is directed at object in order to detect reflected responses from objects. Example-Radar, airbone laser.

b) Passive RS

The sensors detect the naturally reflected or emitted electromagnetic radiation.

2. Based on spectral regions of electromagnetic radiation used

a) Visible and reflective RS

The energy source is the sun. The data obtained by remote sensing depends on the reflectance of objects on ground.

b) Thermal infrared RS

The source of radiant energy is the object itself. Any object at normal temperature (300 ^{0}K) emits electromagnetic radiation with a peak at 10 μm.

3. Microwave RS

In this type, the sensor detect the microwave radiation emitted by an object covering spectral range of 0.1 to 100 cm.

4. Based on sensor platform used

a) Airborne RS

The platforms used for the sensor are flight or unmanned aerial system (UAS) at a desired height.

b) Satellite RS

The carrier of sensor is the satellite which placed in low earth orbit at a height of 160 to 2000 km or Geostationary earth orbit at a height of 30,000 km from earth surface.

5. Based on number of spectral bands

a) Multi spectral RS

The number of bands in multispectral sensors ranges from 3 to 10 having green, red, IR, SWIR etc. Ex- Remote sensing series of India, Landsat of USA and SPOT satellite of France.

b) Hyper spectral RS

Hyper spectral sensors have as many as 200 contiguous spectral bands which measure the entire electromagnetic spectrum.

Ex- Hyper ion of EO- 1 and AVIRIS (Airborne Visible and infrared Imaging spectrometer) of NASA.

Spectral signatures of earth surface features

Every object on Earth's surface has its unique spectral reflectance. Fig. 12.1shows the average spectral reflectance curve for three typical earth featuresi.e. water, vegetation and soil. The spectral reflectance curve for healthy vegetation manifest the **peak and valley** configuration. The valleys in the visible portion of the spectrum indicate pigments in plant leaves. Dips in the reflectance at 0.65 μm, 1.4 μm and 1.9 μm attribute to absorption of water by leaves. The soil curve shows a more regular variation of reflectance.

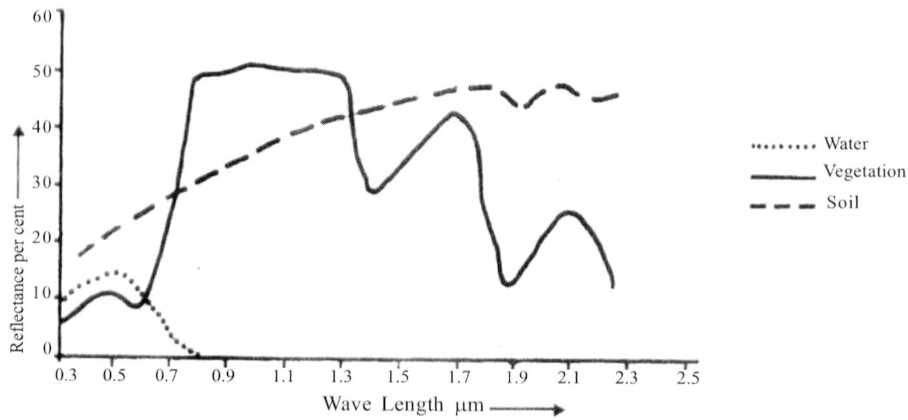

Fig. 12.1. Special signatures of typical features of earth surface

The reflectance of soil curve is affected by soil moisture content, soil texture, presence of organic matter and surface roughness. The most distinctive characteristic of water is the energy absorption at near infrared wave length.

Advantages of Remote Sensing

Remote sensing is used for various purposes such as:

i) Management and exploration of natural resources (oil, mineral)

ii) Planning and land management for highway design, roads, city planning

iii) Resource management in the field of agriculture, forestry, fisheries, hydrology, oceanography and

iv) Atmospheric and weather study.

Use of remote sensing technology helps in several ways such as:

- Large area coverage for survey, identification or classification of features
- Repeated monitoring of dynamic phenomena like crop growth, salinization, alkalization, water logging etc. and
- Data acquisition at multiple heights in different scales and resolution.

Geographical Information System (GIS) and its application

Many development programmes viz. agricultural land use planning, watershed management, reclamation of degraded land/waste lands required precise data on characteristics, classification and potential of soil. All such programmes depend on land evaluation, which many times questionable because of limitation in mapping techniques, lack of suitable models for interpretation of soil and their validation. GIS emerged as an advance tool to overcome the limitation of traditional approach of thematic maps which facilitates integration of the total range of variations and allows visualization of their interactions and spatial relationships.

Definitions of GIS

Several authors define GIS as:

- A Powerful set of tools for collection, storing, retrieving at will, transforming and displaying spatial data from the real world.
- A system for capturing, storing, checking, manipulating, analyzing and displaying data which are spatially referenced to the earth.
- An information technology which stores, analyses and displays both spatial and non-spatial data.

Uses of GIS

GIS can be used as a decision support tool in solving complex environmental problems, forestry and wildlife tracking, watershed managements, agriculture and groundwater resources and management through modeling and mapping in form of 2D and 3D.

Components of GIS

GIS is comprised of four key components:

i) Hardware comprising of computer as procuring unit, input and output devices.

ii) A designated soft wares to store, analyze and display geographic information.

iii) Data-both spatial and non-spatial and

iv) Live ware – i.e. people responsible as interface between software and data using specific methods and protocols for required output.

How GIS works

A GIS stores information about the World as a collection of thematic layers on various themes that can be linked together by geography. A layer contains either an explicit geographic references such as latitude or longitude called special data or an implicit reference such as an address, postal code, road name, census tract name called non-spatial data. These are linked by the software. Both spatial and non-spatial data allow the digital database to be exploited in many more ways than a conventional data base making the maps more meaningful.

GIS data models

GIS work with two different types of geographic models – the vector and roster model. In the vector model, informations about points, lines and polygons are stored in X, Y co-ordinates. The location of a point such as a bore hole can be described by a single co-ordinate whereas, liner features such as road, river can be stored as a collection of co-ordinates. The vector model is used for describing continuous features suchas soil type. Roster models are simple and spatial analysis is easier and faster but, it is voluminous.

Global Positioning System (GPS)

GPS is networking of satellites that transmits, coded information which makes it possible to precisely identify locations on earth. It was originally designated for military applications, but now made available for civilian use.

Components of GPS

A GPS system consists of three segments such as:

i) Space segment-The GPS satellites themselves

ii) The control system

iii) The user segment, which includes both military and civilian users and their GPS equipments.

The whole set of satellite is called **Constellation** and also known as Global Navigation Satellite System (GNSS). There are three globally GNSS such as:

i) NAVSTAR of US (System using timing and ranging)

ii) GLONASS of Russia and

iii) GALILEO of European Union.

Indian Regional Navigation Satellite System (IRNSS) is a regional navigation satellite system designed to provide accurate position information services to users in Indian region extending 1500 KM from its boundary. The control segment consists of Monitor stations, Ground antennas and a Master Control Station (MCS).

The monitor stations track all satellites in view, accumulating ranging data. This information is processed at the MCS to determine satellite orbits and update each satellites navigation message. Updated information is transmitted to each satellite via the Ground Antennas.

The GPS user segment consists of the GPS receivers which convert satellite signals, into position, velocity, and time estimates. Four satellites are required to compute the four dimensions of X, Y, Z (Position) and time. The civilian application of GPS includes surveying, natural resource management.

Applications of geoinformatics in soil resource studies

Soil surveys

Soil survey describes a systematic examination, description, classification and mapping of soil in a given area. In past soil surveys has been carried out through conventional approach which is tedious, time consuming, cost – prohibitive and in accessible in terrain and in mountain topography. Now a days Remote Sensing has been universally accepted as a potential tool for soil survey and resource mapping. Satellite data covers a large area comprised of various landforms, vegetation, and geomorphic processes. Based on the remotely sensed document it is possible to classify terrain phenomenon including land use, relief features and soil conditions.

It also provides informations on drainage pattern, hydrology and land use. In India, efforts were made by ISRO and NBSS & LUP, Nagpur to study the application of multiband photography in the early seventies. Subsequently, digital analysis of airborne remote sensing data in Anantapur district of Andhra Pradesh, enabled delineation of soils up to family level.

Further, soil properties that have been successfully determined by remote sensing includes mineral content, texture, soil water content, soil organic matter content, iron oxide content, albedo, temperature and soil structure. Direct sensing of soil properties requires a remote sensor turned into a specific bands of wave length and computer programme designated to interpret the complex data.

Mapping salt affected soils

Salt affected soils include saline, alkaline and saline-alkaline soil which constitute about 6% of the World land mass. These soils occur in arid, semiarid and coastal areas. Quantification of nature, extent and spatial distribution of salt affected soil is necessary for planning and reclamation programme. The extent of salt affected soil in past was assessed by conventional and Remote Sensing approaches. Using remote sensing technology, the coastal salt affected soils in Odisha was estimated as 2.51 lakh ha by Deptt. ofSoil Conservation Govt. of Odisha during 1981. Recent satellite sensors (Resourcesat-1, Cartosat-2, IKONOS-II and RISAT-2) along with GIS enabled to map salt affected soils in large scale.

However, salt encrustation at surface by capillary movement, water logging condition, salt–tolerance crops present constraints in delineation and mapping of salt affected soil. Several remote sensing indices such as salinity Index (SI) , Normalized Differences Salinity Index (NDSI), Brightness Index (BI), and Normalized Differences Vegetation Index (NDVI) was found better than visual estimation.

Rader derived parameters i.e. back scatting coefficients are used for detecting and mapping salt affected soils. Development of 3 D map through **Hyperspectral**approach combined with active remote sensing would be useful to detect the source of salinity, planning for drainage system in coastal salt affected ecosystem.

Precision Farming (PF)

Precision farming is also called as site specific agriculture. Integration of two advanced tools such as GIS and GPS into a single system that can operated at farm level efficiently. Spatial variability of soil properties from one site to another on a landscape required through examination. Soil properties in one section of a field may require management systems quite different from rest of the same field. In practice, farmers provide optimum inputs over the entire field, even though such systems may provide too much or too little of any input (say fertilizers) in different parts of the field. Such system of management is inefficient, failing to remove the soil constraints in one area of a field, while in another area providing excess inputs may lead to environmental contamination.

Precision farming has three requirements

i) Ability to identify each field location

ii) Ability to capture, analyze and interpret agronomic data over time and

iii) Ability to adjust input use for maximum benefit.

The modern geospatial tools such as Remote Sensing, GIS and GPS meet these requirements and proved to be a potential component of precision farming.

The GIS provides for optimum management on site specific basis within a given field. The data on physical and chemical analysis of soil samples taken on a grid basis can be integrated with digital data from a soil survey base map. Interpretation of these data provide information on soil's ability to provide nutrients, water, air for sustainable management of soil and crop. These data will also indicate soil's susceptibility to erosion and contamination. GIS is useful to create fertility, weed and pest intensity maps which can be used for application of nutrients and pesticides.

The most common use of GPS in agriculture is for yield mapping and variable rate fertilizer / pesticide application. Integration of yield data from GPS with soil properties obtained through GIS develop a site specific management plan.

Based on the appropriate tools and techniques, precision farming precisely monitor and assess the agricultural systems at a local and farm level.

Watershed management

Watershed is a geo hydrological region having a common drainage point. The precipitation from a watershed area flows through a single point into a stream. Remote Sensing and GIS play a vital role in watershed management. The entire watershed area can be mapped for various geo spatial data bases using Remote Sensing and GIS techniques. These data base can be converted to generate information on cropping pattern, water harvesting structure, run off estimation and soil loss study. Digital elevation model (DEM) derived from Stereo satellite data such as CARTO SAT or TERRAASTER provide direction of flow, flow accumulation and flow length maps. Integration of these data with Geo Spatial data bases (from GIS) determine suitable sites for conservation measures in a watershed.

Soil erosion by water leads to siltation of major reservoirs in India. Multi temporal satellite data have been used to estimate the silt load in reservoirs.

Estimation of soil properties

Based on the soil properties, the soil surface reflects various wavelength of radiant energy. Depending on the wave length, a number of soil properties can be directly sensed and recorded. Soil properties that have been successfully determined by Remote Sensing includes mineral content, texture, soil water content, soil organic matter content, albedo, temperature and soil structure. Spectral reflectance's in visible and near infrared (VIS / NIR, 400-2500 nm) as well as in mid infrared (MIR) spectral regions have been widely used for assessment of different soil properties (Table 12. 1).

Hyperspectral Remote Sensing has been found superior to conventional broadband multispectral Remote Sensing. However, geostatistical approaches based on spatial dependency have been extensively studied in GIS environment for soil parameter prediction beyond sampling location considering spectral measurements as primary source. Different soil attribute maps are generated from point measurements using different geo statistical methods.

Table 12.1. Selective spectral bands in a satellite-based remote sensors and its field application.

Spectral Region	Spectral bands	Field application
VIS (Visible)	0.4—0.7μm	Surface chemical composition, vegetation
NIR (Near infrared)	0.7-1.1 μm	Biological properties
SWIR (Soft wave infrared)	1.1 – 1.35 μm	Surface chemical composition and
	1.4 – 1.8 μm	atmospheric chemical composition
	2.0-2.5 μm	
MWIR (Mid wave infrared)	3.0-4.0 μm	Surface chemical composition and
	4.5-5.0 μm.	atmospheric chemical composition
Microwave	0.1-100 cm	Surface heat capacity, temperature, surface & atmosphereic constituents.
TIR (Thermal infrared)	8.0-9.5 μm 10.0 -14μm	Surface physical properties, atmospheric precipitation, Surface temperature,

Study Questions

1. Explain potential use of Remote Sensing, GIS and GPS in precision farming.
2. Describe different types of Remote Sensing based on its energy source, spectral range, and platforms.
3. Explain or define the following:

i) Geo informatics

ii) Remote Sensing

iii) GIS

iv) GPS

4. Explain the principles and basic processes of Remote Sensing.
5. Explain different data models used in GIS.
6. Explain the working principle of GPS. What are different GNSS available globally?
7. Briefly discuss the application of Remote Sensing and GIS in mapping of salt affected soils.
8. How can the GPS be used in GIS?

13

Land Use Planning

Sustainable management of soil and water resources is essential for food, nutritional and environmental security of the country. The total and partial factor productivity of agricultures in India declined over years because of continued degradation of natural resources. The process leading to land degradation are generally triggered by excessive use of land and water resources to meet the growing demand of population for food, fibre and fodder. Various anthropogenic activities leading to accelerated land degradation through salinization, flooding, water logging, erosion of water and nutrients, drought reduced agricultural productivity and nutritional security in the country. These problems can be overcome by optimum utilization of land adopting suitable management practices. For effective land management practices, the soils should be interpreted in forms of their land capability, suitability for irrigation and crop growth. This will not only help the farming community and extension specialists, but also to the administrators and policy makers for using soils for sustainable agricultural production.

Land Capability Classification

The purpose of land capability classification is to group land units, into defined classes based on their capability. It serves as a guide to assess the suitability of land units for arable crops. grazzing / forestry. The **objectives**of land capability classifications are:

i) Use the land according to its capability,

ii) Treat the land according to its needs and

iii) Produce from the land according to its potentiality.

The factors which determine the capability of land are:

i) The inherent characteristics of soils such as morphological, physical and chemical properties of soils (Soil depth, texture, structure, permeability, fertility status.).

ii) External land features such as slope, extent of erosion, susceptibility to over flow, flooding, water logging, stoniness etc.
iii) Environmental factors such as rainfall, temperature and
iv) Present land use.

The capability classification consists of three categories namely

i) Land capability units
ii) Land capability sub class
iii) Land capability class

i) Land capability units

It is a grouping of one or more soil mapping units having uniform soil characteristics, potential and limitations and require fairly uniform conservation treatment and management. The land capability units is expressed by putting an Arabic number (1, 2, 3) at the end of land capability class and sub class. Example $IIIe_1$.

ii) Land Capability Subclass

Sub classes are the groups of capability units having same kinds of dominant limitations such as wetness or excess water (W), erosion and run off (e), root zone or soil limitations (s) and climatic limitations (c). The suffixes e, w, s or c are added to the capability class symbol as: IIe, IIIw, IIIc etc.

iii) Land capability class

Land capability classes are the groups of land capability units and land capability sub classes having same relative degrees of limitations and hazards and are indicated by Roman numerals such as I, II, III etc. The land capability classes broadly divided into two groups as class I to IV and V to VIII.

The limitations in land use and management practices increases from class 1 to class VIII (Table 1).

Table 1. Structure of land capability classification.

Capacity	Class	Subclass	Units
Arable	I		
	II	IIe erosion	IIe1
	III	IIw wetness	IIe2
	IV		
	V		
Nonarable	VI	IIs, soil	IIe3 etc.
	VII	IIc climate	
	VIII	IIes erosion and soil	

A brief description of the characteristics and safe use of land and soil of each class is described in following paragraphs

Class I Very good cultivable land, deep, leveled, well drained with almost no limitations. The soils are suitable for a variety of crops including wheat, cotton, maize, vegetables etc.

Class II Soils in this class have slight limitations of soil depth, salinity, drainage or erosion that reduce the choice of crops. These soil are suitable for wheat, barley, cotton and moderately suitable for maize and vegetables.

Class III Moderately good cultivable land on level plain or gently slopes having severe limitations. Limitations in choice of crops due to moderate slope, high erosion hazards, moderate depth and restricted root zone ,low fertility, low water holding capacity, moderate salinity or alkalinity, unstable soil structure. Unsuitable for vegetable crops. Cover crops such as legume, grasses must find a place in crop rotation. Class III soil often requires special conservation practices.

Class IV Can be used for cultivation but, have severe limitations on the choice of crops due to steep slope, severe erosion, shallow soils, poor drainage, low water holding capacity, severe salinity and alkalinity.

These soil may not be economical for cultivation as they need frequent soil conservation and management practices.

Class V These types of lands are not suitable for arable cropping but, suitable for grazing. Frequent over flow, stony or rocky soils, water logging because of poor drainage pose severe limitations in land use.

Class VI Non arable land, well suited for pasture or range, woodland or wildlife. The physical limitations are same as those for class IV but, degree of severity is more.

Class VII Have severe limitations which restricts their use for grazing, wood land or wild life. Need careful management for grazing and pasture.

Class VIII These soils are suitable for wildlife or recreation because of severe limitations. The lands included in this class are tidal lands, swamps, river wash, sand dunes and barren mountain tops etc. Various features of land capability classes are listed in Table 2.

Table 2. Land capability classification criteria.

Characteristics	Class I	Class II	Class III	Class IV	Class V	Class VI	Class VII	Class VIII
Topography (t)								
Slope (%)	0-1	1-3	3-8	8-15	Upto 30	15-35	35-50	>50
Erosion	Nil	Slight	Moderate	Severe	Nil	Severe	V. Severe	-
Wetness (w)								
Flooding	Nil	Nil	Nil to slight	Slight to moderate	Moderate to severe	Nil to severe	Nil	Excessive
Drainage	Well	Moderate	Imperfect	Poor	V. Poor	Excessive	Excessive	-
Permeability	Moderate	Moderate	Rapid – slow	V. Rapid – V. Slow	-	-	-	-
Infiltration rate (cm/h)	2 – 3.5	1-2	0.5-1.0	<0.5	2..0	-	-	
		3-5	5.0-10.0	>10.0				
Surface Physical conditions (s)								
Texture	Loam	Sil&Cl	Sl& C	SCl	SC	LS, Cl	LS,S,C	LS,S,C
Coarse fragment (Vol %)	1-3	3-15	15-40	40-75	15-75	75+	-	-
Stoniness (%)	<1	1-3	3-5	5-8	8-15	18-40	40-75	**>75**
Coarse fragments (%)	<15	< 15	15-35	35-50	50-75	50-75	50-75	>75
Soil depth (cm)	>150	150-100	100-50	50-25	-	25-10	25-10	<10
Profile Development	Cambic/ Agrillic horizon A-B-C	A-B-C	Stratified A-C,A-B-C	Salic / calcic horizon A-BZ-C/ A-BK-C	AZ-C A1-B2-C	Gypsic horizon A-CY	A-C (Stony)	A-C (Boundary)
Fertility								
CEC(c mol (+) kg^{-1})	40-16	16-12	16-12	-	-	-	-	-
Base saturation(%)	80+	80+	80-50	50.-35	50-35	35-35	<15	-
O.C (%)	>1.0	0.75-1.0	0.5-0.75	<0.5	<0.5	-	-	-
Salinity EC (dS m^{-1})	<1.0	1-2	2-4	4-8	8-15	15-35	35-50	>50
Gypsum (%)	0.3-2.0	2-5	5-10	10-15	15-25	>25	-	-

Land irrigability classification

Classification of land for irrigation purpose is based on three factors such as ;

i) Drainage of land,

ii) Effect of irrigation water on soil salinity and alkalinity and

iii) Cost of land development.

Based on the limitation of different land characteristics such as texture, depth, permeability, salinity and topography, FAO soil scientists along with Iranian workers classified the land into six classes as –

Class 1 to 3	Arable, suitable for irrigation having nil to moderate limitations
Class - 4	Restricted arable with severe limitations of soil
Class- 5	Undermined arable having severe limitations of soil salinity and/or drainage for irrigated agriculture
Class -6	Non arable, unsuitable for irrigation

In order to quantify the irrigability approach a parametric approach for evaluating the suitability (capability index Ci) for irrigation was proposed by **Sys (1976)** considering seven parameters as:

Soil physical and chemical properties - Texture, available WHC, salinity, CaCO3 and gypsum content

Drainage Condition – As deduced from internal drainage, permeability and intake rates

Topographical condition - As influenced by slope

Accordingly the rating to each soil unit is evaluated on the degree of limitation (s) (Table 3).

The overall capability index (Ci) is calculated as:

Capability Index (Ci) = A. B. C. D. E. FWhere,

A - Rating for soil texture - taken as 100 for best texture, say loam

B - Rating for soil depth

C - Rating for $CaCO_3$ status

D - Rating for gypsum status

E - Rating for drainage condition

F - Rating for topography (slope)

(B to F: Taken as faction of 1)

On the basis of Ci, the soil suitability class of mapping unit is determined. The following classes and sub classes are defined.

Classes	Suitability	Ci rating
S1	Very suitable to suitable	>80
S2	Moderately suitable	60-80
S3	Marginally/slightly suitable	40-60
N1	All most unsuitable	20-40
N2	Unsuitable	<20

Sub classes

S - limitations of physical soil properties (A, B, C, D)

N - limitations due to salinity / alkalinity (E)

D - drainage or wetness limitation (F)

T - topographical limitation (G)

Example for Calculation of Ci

Calculate the Ci of a soil having loam texture (A), very deep (B), 5-10% $CaCO_3$ (C), 2-5% gypsum (D), salinity 4-8 dS m^{-1} (E), well drained (F) and have slope < 1 % (G).

Capability Index (Ci) = A. B. C. D.E.F.G

= 100 x 1.0 x 0.9 x 0.8 x 0.9 x 1.0 x 1.0 = 64.8 or say 65.0

Table 3. Rating based on kind and degree of limitations for surface irrigation.

Land Characteristics	Land class, degree of limitations and rating					
Class		S1	S2	S3	N1	N2
Degree of Limitations	0	1	2	3	4	5
Rating		90	80	60	40	20
A. Texture	Clay loam	L, Sil, Cl, SiCl	Fine – L, Fine –C SCL	V.F-C LS	Fine sand	Coarse sand
B. Soil depth (cm)						
Soft rock	150+	100-150	75-100	50-75	50-20	<20
Hard rock	300+	200-300	100-200	50-100	50-20	>20
C. CaCO3 status (%)	1-10	10-25	25-40	40-75	>75	-
(as per Sys)	0-5	5-15	15-30	30-50	50+	-
D. Gypsum status (%)	<0.5	0.5-2	2-5	5-10	10-25	>25
E. Salinity /alkalinity						
EC (dSm^{-1})	<4.0	4-8	16-30	30+		
ESP	<5.0	5-15	15-25	25-45	45+	

F. Drainage

Internal drainage	Moderate Slow		Slow to rapid	Very slow to rapid	Very slow to rapid	Very slow
Infiltration (cm h^{-1})	3.5 – 7	0.8-3.5	0.2-0.5 7-10	0.1-0.2 10-25	0 - 2 2-5	
Natural drainage	Well	Moderate	Imperfect	Poor	V.poor	V.poor
G.Topography (slope %)	0-2	2-5	5-8	8-15	15-25	>25

Source: Sehgal (2014)

Land Suitability Classification

Land suitability classification is the process of land evaluation for specific kind of land use such as rainfed agriculture, forestry, fodder etc.

The availability of plant nutrients and water is largely controlled by the physico -chemical properties and micro environment of soils. The success or failure of plant species in a particular area is largely governed by these factors. The deep rooted forest plantation crops respond differently to soil texture, soil depth than the shallow rooted arable crops such as wheat, rice, maize etc.

Structure of Land Suitability Classification

The FAO (1976) panel for land evaluation suggested the classification of land for specific use. There are four levels of classification as orders, classes, subclasses and units.

1. Land suitability orders

Land suitability orders indicate whether the land is suitable or not suitable for use under consideration. There are two orders, represented by symbol S and N.

Order S, suitable - Land on which sustained use of the kind under consideration is expected to yield benefits which justify the inputs, without unacceptable risk of damage to land resources.

Order N, not suitable - Land which has qualities that preclude sustained use of the kind under consideration.

2. Land Suitability Classes

Land suitability classes reflect the degrees of suitability. These classes are numbered by Arabic numbers such as S1, S2, N1, or N2. Different land suitability classes within an order are given below:

Class S1 (highly suitable)

Land having no significant limitations to sustained application of a given use or only minor limitations that will not significantly reduce benefits and will not raise inputs above an acceptable level.

Class S2 (moderately suitable)

Land having limitations which in aggregate are moderately severe for sustained application of a given use. The limitations will reduce productivity of benefit and increase required inputs to the extent that the overall advantages to be gained from the use.

Class S3 (Marginally suitable)

Land having limitations which in aggregate are severe for sustained application of a given use and will so reduce productivity or benefits or, increased required inputs that this expenditure will be only marginally justified.

Class N1 (Currently not suitable)

Land having limitations which may be surmountable in time but which cannot be corrected with existing knowledge at currently acceptable cost.

Class N2 (Permanently not suitable)

Land having limitations which appears so severe as to preclude any possibilities of successful sustained use of the land in the given manner.

3. Land suitability sub classes

Land suitability sub classes reflect the kinds of limitations or main kind of improvement measures required within a class.

The following sub classes have been defined.

c - climate

t - topography

w - wetness

n - salinity / sodicity

f - soil fertility

s - physical (influencing soil / water relationship)

Land Suitability Units

Land suitability units are divisions of sub classes that differ from each other in detailed aspects of their production characteristics or management requirements. They are indicated by an Arabic numerals following subclass symbol. For example – S2 d-1, S2d-2 which indicates there might be two types of lands assigned to S2 on ground drainage.

Summary

The structure of land suitability classification is summarized and given below:

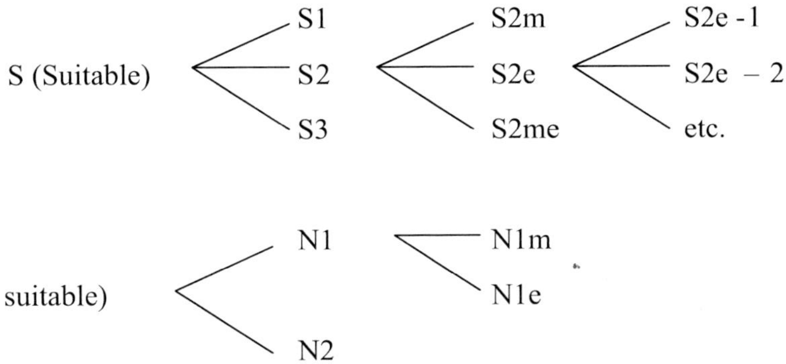

Sys and Debayeva (1975) proposed the capability index (Ci) based on nine parameters in the arid and semiarid regions.

Ci - A.B.C.D.E.F.G.H.Iwheres

Ci - Capability Index

A - Soil texture

B - Calcium carbonate

C - Gypsum

D - Salinity

E - Sodium saturation

F - Drainage

G - Soil depth

H - Epipedon

I - Profile development

Sys. (1976) proposed the following scheme for evaluating the degree of limitation ranging from 0 (no limitation and having Ci> 80) to 4 (Severe limitation with Ci of 30 or less).

0 **No limitation** - Ci 80 or more; quality optimal for plant growth.

1 **Slight limitation**- Ci 60-80, the characteristics are nearly optimal for land utilization type. The productivity only decrease by 20% from optimum yield.

2 **Moderate limitation**-Ci 45-60; the characteristics have moderate influence on crop yield decline.

3 **Severe limitation**-Ci 30-45, the characteristics have severe effect on productivity and the land use became uneconomical for the considered land utilization type.

4 **Very severe limitation**-Ci Less than 30;limitation will decrease yield below the profitable level and also inhibit the use of soil for the considered land utilization type.

A Schematic limitation is given below;

Limitation Level	Class levels
0	S1
1	S1
2	S2
3	S3
4	N 1 and N 2

The limitation approach has been successfully used to provide a qualitative land evaluation based on soil properties which are available after a quality soil survey of an area.

The said quality parameters for evaluating land for agriculture, forestry and plantation crops for defining suitability classes are:

Soil site characteristic	Related land quality
Climate (c)	Available moisture
Topography (t)	Resistance to erosion
Wetness (w)	Available moisture
Drainage, flooding	
Physical Condition (S) of soil	
Texture	Water availability
Stoniness	Root development
Depth	Plant growth
Calcium carbonate	Nutrient availability
Gypsum	Source of sulphur
Soil fertility (f) (not readily correctable)	
Organic matter	Source of nutrient and physical condition
CEC	nutrient holding capacity
Base saturation	Nutrient availability
Nutrient availability	Soil fertility
Salinity and alkalinity (m)	
Salinity	Excessive salts
Ground water depths	Salinity and water logging
Alkalinity/sodicity	Source of sodicity

References

FAO (1976) A Frame Work for land evaluations. Soils Bull, FAO, Rome.

Sehgal, J. (2014) A Text Book of Pedology – Concepts and Applications: pp. 353.
Sys, C. (1976) Land evaluation. ITC, State University, Ghent, Belgium.

Sys, C. and Debayeva, J. (1975) Land evaluation. Vol. 1 Agric Publ.7. Gen. Adm. For Dev. Coop. Brussels (Belgium).

Study Questions

1. State the objectives of land capability classification and factors affecting it.
2. The land capability classification of a field in a watershed is given by IIIe2. State the land capability class, sub-class and unit of the land.
3. Differentiate between land capability Class III and Class VI with reference to soil conservation and management practices.
4. Briefly discuss different features of land irrigability classes.
5. Name different parameters for calculation of land capability index (Ci) for irrigation.
6. What do you mean by land suitability classification? How it is different from land capability classification.
7. State the structure of land suitability classification with an example.
8. Name different parameters to calculate capability index (Ci) for land suitability for crops.
9. "Land capability classification is strongly biased towards consideration of soil conservation". Justify the statement.
10. What is the necessity of land use planning in present scenario?

14

Multipurpose Tree Species in Problem Soils

Trees are an integral and indispensable constituents of our natural surroundings. The trees keep the environment clean, make the soil healthy and provide food, fibre and shelter to mankind and wild animals. Trees help in stabilizing the soil, a vital natural resource, prevention of soil and water erosion, control run off water in the catchment areas, acts as a shelter belt against wind erosion in arid zone, act as a wind breaker to protect the coastal region during storm and cyclone besides contributing substantially to economic development. The National Commission on Agriculture stress the need for afforestation of about 40 million ha of land presently lying barren because of salinity, sodicity, waterloggings, ravines, coastal sandy areas, stony and gravelly lands, high altitude and slopping lands. Several agencies in various parts of the country are engaged in afforestation of these degraded lands.

Multipurpose trees – their role in waste land management

Definition

Multipurpose trees (MPT) are defined as all woody perennials that are purposefully grown to provide more than one significant contribution to the production and / or service functions of a land use system.

Benefits derived from MPT

1. Supply food in form of fruits, nuts or leaves that can be used as a vegetable.
2. Supply fire wood, timber and fodder.
3. Add nitrogen through nitrogen fixation in soil.
4. Act as a wind breaker in coastal ecosystem and shelter belt in arid zone.
5. Used as a living fence in home garden.
6. Find ways of making use of marginal lands sustainable, reclaim and restore degraded land.

7. Decomposition of leaves and biomass maintain soil organic matter, biological activity and improve soil structure, water retention capacity.
8. Used as an insecticide in Integrated Pest management Programme (IPM), act as nitrification inhibitor (neem coated urea). Ex. Neem, Karanj.
9. Improve nutrient cycling and nutrient use efficiency.

Ideal Characteristics of MPT

Selection of tree species to obtain maximum benefit depends upon the plant characteristics such as;

- High rate of leafy biomass production
- Existence of deep roots
- A high rate of nitrogen fixation
- Spreading habit to be used as cover crops
- Absence of toxic substances in the litter or root residues
- Capacity to grow in adverse soil conditions (water logging, salinity, sodicity, acidity)
- Should not have any adverse effects on adjacent crops

For an example, planting of Eucalyptus spp. in water logged soils lower the water level and reduce salinization, but adversely affects the growth and yields of adjacent crops.

MPT for sodic soils

The salt affected soils occur extensively in different agro ecological and soil zones of the country, particularly in the arid, semiarid and sub humid regions. Out of 7.5 m ha of salt affected soils, about 2.5 million ha are alkali soil. Alkali soil are characterized by high pH |(>8.2), low EC (< 4.0 dSm^{-1}) and high ESP (>15%). Presence of high exchangeable sodium in exchange complex adversely modified the growth of most plants. Poor soil structure with low infiltration rate leads to excessive run off. Growing of trees in barren alkali land appears to be a promising land use. Afforestation in these barren lands would also contributes to the maintenance of healthy agro-ecological system and averting environmental hazards.

A successful plantation programme depends upon

i) Good survival and rapid growth of saplings

ii) Choice of right species matching the planting site

iii) Adopting a good planting techniques, fertilizer application, weed control and irrigation during initial stage.

Tree species in sodic soils

Among the tree species tried by CSSRI, Karnal, the promising species suitable for highly alkali soils are: *Acacia nilotica, Prosopis juliflora Casuarina equiseti folia* and *Sesbania aegyptiaca.*

The moderately good spp. are *Albizia lebbeck* (Siris) and *Azadirachta indica* (Neem). *Eucalyptus terreticornis* performed well during initial 2-3 years but, later the growth was restricted. Visual symptoms like yellowing or chlorosis indicated possibility of nutritional disorders.

The trees were planted in 120 to 150 cm deep auger holes, filled with 7-8 kg FYM, 2-3 kg gypsum and alkali soil from the planting site. These trees can yield 5-10 t/ha biomass annually on pruning after 2 and 4 years of growth. The biomass can be used as fodder and organic matter.

MPT for saline soils

The saline soils occur in arid and semiarid regions and coastal eco- system. The soils are rich in soluble salts having EC > 4 dSm^{-1}, pH neutral to 8.0 and ESP < 15%. Vast area in coastal ecosystem remained barren or fallow because of high salinity during post monsoon period. Afforestation in these lands would provide food and fire wood and also act as wind breaker against storm and cyclone.

Suitable tree spp. for saline soils include *Prosopis juliflora., Acacia nilotica, Porospis chinensis.* The leaf biomass of Prosopis spp. can be used as fodder.

Coconut (*Cocos nucifera*) and arecanut (*Areca catechu*) are good plantation crops grown successfully in coastal ecosystem in states of Keral, AP, Odisha and West Bengal. These trees grow well under saline soils rich in sodium and potassium. These crops are grown for food, fuel and timber.

The varieties of fruits and vegetables suited to arid, semi-arid saline soils are given in Table 1.

Table 1. Relative tolerance of fruit crops to salinity.

Tolerant- EC >6 - 8 dSm^{-1}	Moderately tolerant- EC 6 -3 dSm^{-1}	Sensitive-EC- 3-1.5 dSm^{-1}
Datepalm	Fig	Peach
Ber	Orange	Apricot
Pomogranate	Lemon	Avocado
Custard apple	Mango	Almond
Kair Guava	Grape	Plum

Jhau, casurnia grow well in coastal tract which act as wind breaker against storm and cyclone. Over the last three decades the Govt. of Odisha has been planting Jhau (*Casuarina equisetifolia*) trees along the coastal areas of Odisha as a part of coastal afforestation initiative to protect coastal ecosystem. *Pandanus fascicularis*, a saline resistant plant is seen in swampy areas behind Gopalpur sand beach in Ganjam district, Odisha. It protect the sea coast from high tides. Its flower having beautiful smell is used in perfume industry.

Mangroves, the salt tolerant forest ecosystem comprises of plant species at the interface of land and sea in sub-tropical inter-tidal regions of the world. Important mangrove plant species includes Bani, Guan, Keruan, Sundari, Orua, Hentala in BhitarKanika, Odisha.

The forest is locally known as **Luna Jungle** or **Hental Jungle**. The forest species have a high tolerance to saline environment.

Mangrove plants are the source of rich food for the organisms of the mangrove ecosystem. Mangrove forest is a natural habitat for salt water crocodiles, reptiles, birds like king fisher, pelican and mammalian species including spotted deer, sambar, fishing cat, otter etc.

MPT for acidic waste lands

Acid soils are characterized by low soil pH, low availability of macro, secondary and micronutrients such as boron and molybdenum. In rolling topography, the acidic red and laterite soils remained barren due to loss of top soil through run off water.

Besides, various soil and water conservation measures, various types of plantations including energy plantation, fuel and fodder species, dry land fruits can be taken up in these areas depending upon the soil capability and climatic conditions. These species not only protect the land, but will produce food, fodder, fuel and timber.

With people's participation, the Deptt. of Social Forestry recommended following plantation programmes for the benefit of villagers-

1. Government waste lands and community land plantation
2. Degraded barren hill planting
3. Rehabilitation of degraded forests
4. Industrial plantations
5. Farm forestry and
6. Dryland horticulture

Different tree species suited to degraded acid soils are listed in Table 2.

Table 2. List of tree species for degraded acid soils

S.No.	Trees
1.	Harida (*Terminalla chebula*)
2.	Mango (*Magnifera indica*)
3.	Tentuli (*Tarmarindus indica*)
4.	Mohua (*Madhuca latifolia*)
5.	Piasala (*Pterocarpus marsupium*)
6.	Bahada (*Termina liabellirica*)
7.	Jamu (*Eugenia jambolana*)
8.	Simili (*Bombax malabaricum*)
9.	Karanja (*Pongamia glabra*)
10.	Sal (*Shorea robusta*)
11.	Gambhari (*Gemilina arborea*)
12.	Amla (*Emblica officinalis*)
13.	Mulberry (*Morus alba*)
Horticultural species	
1.	Ber (*Ziziphus mauriitiana*)
2.	Custard apple (*Annona squamous*)
3.	Mango (*Mangifera indica*)
4.	Guava (*Psidium Guajava*)

MPT for waterlogged and wetlands

Waterlogged wet lands are found in swampy and marshy areas where soil is badly or imperfectly drained. Vast areas of fertile land have gone unproductive due to rise in surface water table and accumulation of salts and toxic elements. Tree species with high tolerance to salts and water, fast growing habit are good for water logged saline and alkali soil. Acacia nilotica, *Prosopis juliflora*, *Tarmarix articulata* have been found suitable for these lands. In wet soils Eucalyptus robusta, *Salix* spp., *Populus nigra*, Acacia nilotica gave better results. For marshy areas, *Barringtonia* spp. *Eucalyptus robusta*, *Casurina equisetifolia* have been found successful.

MPT for stabilization of sand dunes

Stabilization of sand dunes depends on reducing wind velocity at the surface through wind breaks or establishing vegetative cover.

Afforastation with tree spp like *Prosopis cineraria, P. juliflora, Acacia nilotica, Acacia senegal, Acacia tortilis* are found suitable to stabilize sand dunes.

Study Questions

1. Define MPT. State different criteria for selection of MPT.
2. What are beneficial effects of MPT in managing degraded land?
3. Name different MPT species suitable for salt affected soils.
4. Name different plant species found in a mangrove forest. How do these plant spp. provide shelter to organisms in a mangrove ecosystem?
5. What are the characteristics of tree species suitable for
   a) Dry land region
   b) Water logged areas
   c) Stabilization of sand dunes.
6. State the beneficial and harmful effects of Eucalyptus on Water tables and adjacent crops.
7. Name some tolerant, moderately tolerant and sensitive fruit crops to soil salinity.

15

Land Management in Special Problem Areas

Soil conservation in hilly areas

Soil erosion is a major problem on hills because of steep slopes. It has been estimated that about 6 million ha of land requires intensive conservation measures in Himalayan and sub-Himalayan regions of Uttar Pradesh, Uttarakhand, Punjab and Himachal Pradesh. Anthropogenic activities like shifting cultivation, deforestation, over grazing accelerate soil erosion in hilly ecosystem.

Soil conservation in hilly areas requires a well-planned and rational land use programme such as:

i) Steep slopes may be put under permanent forest,

ii) Gentle slopes should be terraced and

iii) Valley bottom should be put under suitable agricultural crops.

This can be achieved by treating the entire areas on a complete watershed basis for rational use of forestry, agriculture and horticulture. A combination of engineering measures such as terracing along with afforestation, torrent correction and silvipastoral practices will reduce the erosion on hills.

Management Practices

a) Afforestation

Forests play an important role in soil conservation. Afforestation of economic plants should be done on contour benches. The suitable plant species for dry areas of north–west India are *Acacia modesta* and *Prosopis juliflora.* For cooler areas of Himachal Pradesh, Leh and Kargil in Jammu and Kashmir *Populus alba*, *P. nigra, P. Cilita* are the most promising ones. Indiscriminate grazing by cattle, sheep and goat are major causes of soil erosion. It is necessary, that pastures and grasslands in hilly areas should properly be managed for maximum fodder production. It will maintain soil fertility and control erosion.

b) Torrent and slip correction

In northern and north eastern Himalayan region, high run off down the hill slopes with heavy silt load fill up the channel beds and reduce their capacity to carry run off and sediments. Land slips can be treated by providing proper drainage using check dams with vegetative covers.

c) Terracing

Graded bench terracing is recommended for Himalayan regions of Punjab, Himachal Pradesh with slope ranging from 16-33% and graded trenches on 10-15 % slopes. Eroded lands can be protected with soil binding cover like grasses.

d) Management of gullied land

Small gullies (3 m deep) can be reclaimed by clearing, minor leveling and construction of a series of contour bunds at a vertical interval of about 0.9m.

A medium gully (3-9m deep) can be reclaimed by clearing, leveling the gully and constructing a series of earthen and brick masonery check dams.

The best land used for deep and narrow glullies is permanent vegetation. Gully beds could be protected with live check dams. Gully erosion can be controlled by raising perennials like **Acacia nilotica** and grasses.

e) Land management and crop planning

i) Hill tops and steep slopes should permanently be placed under forest species.

ii) Lower down forestry should be place to Jhum cultivation. Gentle slopes should be terraced and valley bottoms should be placed under agriculture. Horticulture should be developed in suitable localities. Cultivation of fruits such as oranges, pineapple, Jack fruits, banana, papaya are recommended. Crops such as potato, tapioca, turmeric and ginger have potential for these areas.

iii) The fuel and fodder plantations is probably the best system to replace Jhuming.

iv) Conservation agronomy practices have been found to be simple and economical for slopes ranging from 0.5 to 6.0 percent. Contour cultivation, contour farming, inter-cropping, minimum tillage, strip cropping and bunding, ridge and furrow method of planting can be practiced to conserve soil and water.

v) Water harvesting, storage and recycling - The excess run off stored in farm ponds can be used for life saving irrigation.

f) Waterlogged and wet lands

Waterlogged and wet lands include seasonal or permanent waterlogged marshy and swampy areas. The total area of all kinds of wet land in the country is about 20 million ha. Land drainage is essential for management of these soils. Drainage provides several advantages such as permits aeration which is essential for root penetration and growth, growth of beneficial micro-organisms, remove toxic substances from the root rhizosphere etc.

Depending upon land condition, surface and subsurface drainage is recommended. Digging a pond in 20% area of the field will be helpful in several ways. The store water can be used for life saving irrigation and fish cultivation. Fruit crops like banana, papaya, coconut etc. can be planted on raised bunds around the pond.

Soil Management in Arid Zone

In arid zone, about 18.79 m ha is occupied by both hot and cold desert spread over in the states of Rajasthan, Gujarat and small pocket in Punjab and Haryana. In arid zone, the moisture deficit index is less than or equal to -40, with annual rainfall ranging from 300 to 500 mm and mean PET of 1400 to 1700 mm. The landforms comprise of rocky stretches, shifting dunes, sandy plains, saline flats, alluvial plains, dead streams, barren and undulating topography.

Problems of Arid Zone of India

Deforestation and overgrazing leads to land degradation by wind erosion. The causes of wind erosion in the desert region are, low rainfall, high wind velocity and sandy soils with loose single grained structure, faulty method of land use, over grazing and deforestation.

Management

In arid zone, class VI lands predominate and account for 45% of total area which are highly susceptible to erosion. These areas should be put under permanent pasture. Rocky pediments should also brought under permanent silvi – pastoral cover with light grazing intensity.

Water is the key challenge for food production in arid region due to variability of rainfall, long dry seasons and dry spells. Water conservation measures including in situ moisture conservation, mechanical measures such as bunds, drainage line treatments, water harvesting and re-use and ground water recharge are recommended.

Bajra is the predominant crop in arid zones followed by cluster beans and sesame. Gram and peas are generally grown in winter season. Field strip cropping of mungbean in grass strip was found successful for control of wind erosion. The marginal and sub marginal lands should be brought under grassland to curb erosion.

Sand dune stabilization and shelter belt plantation are two major activities to control wind erosion.

Sand dune stabilization programmes include activities such as protection of area from human and livestock encroachment, creation of micro-wind breaks on sand dune slopes using locally available shrubs, plantation of grass slips or direct sowing of grass seeds on leeward side of micro wind breaks. It has been observed that for sand stabilization, small shrubs and grasses are better sand binders than the tree species.

Erection of shelterbelts along the boundaries of crop fields helps to reduce injuries to the tender seedling from sand blasting and hot desiccating winds. Establishment of micro shelterbelts in arable lands by planting tall and fast growing plant species like castor on the wind ward side and shorter plants like vegetable crops in the leeward side of tall plants helped to increase the yield of crops. Planting of trees on field bunds across the direction of wind are suggested.

Water logging and salinization pose another challenge to the arid environment. Bio-drainage by plantations of Eucalyptus and Accacia around the water logged area was found effective. It was reported that these spp. transpired enough water to lower the water table by 15 m over a period of 6-7 years.

Management of problem soils in semi-arid zones

The semi- arid areas cover one-third of India spread over north-eastern and western Madhya Pradesh, Marathwada and Vidarbha regions of Maharashtra, Telangana region, Saurashtra region, Tamil Nadu, Karnataka, Southern Rajasthan, eastern Haryana and western Utter Pradesh. The annual rainfall varies between 500 to 900 mm and moisture deficit index of -40 to -60. The moisture deficit index is given by-

$$\text{Moisture deficit Index} = \frac{P - PE}{PE} \times 100$$

Where, P – precipitation and PE – potential evapo transpiration.

In semi-arid zone, the rainfall is short, erratic and highly variable with time and space. Much of rain occurs with high intensity resulting in run off and severe erosion.

Black, red and alluvial soils are the predominant soil types in semi- arid zone. The soil are light texture, low in organic matter, fertility. The crop suffers due to scarcity of water, frequent and prolonged drought. Most of the dry lands are undulating with variable slope both in magnitude and direction.

Soil and Water Erosion

In semi-arid areas, 20-30% of rainfall is lost as run-off leading to soil loss varying from 15 to 81 tonnes/ha. Low rate of infiltration of black soils at saturation resulted in high run off that cause soil erosion. Red and alluvial soils have comparatively high rates of infiltration, but become very susceptible to erosion once run off occurs. Maximum soil loss occurs when the land is fallow without any cover. Cultivation of crops like Jowar, tobacco and maize cause more soil and water erosion as compared to close growing legumes such as cowpea, mung, groundnut.

Soil Management Practices in Semi- arid Zones

The strategies of erosion management on arable lands are discussed in this section.

1. Efficient use of rain water

In semi-arid zones, the rain water can be efficiently managed by:

i) Increasing infiltration capacity of soil and reducing run off,

ii) Water harvesting and recycling,

iii) Reducing the loss of water through evapotranspiration and

iv) Efficient use of water and nutrients.

Increasing Infiltration Capacity of Soil

The moisture content of soil can be increased by allowing more water to infiltrate to the sub-soil region by leveling the land, bunding, deep ploughing during summer season, and mulching.

The beneficial effects of soil moisture storage in the profile through land leveling and bunding has been reported in several parts of the country. Soils with more than 6% slopes are highly susceptible to erosion. Such soils should be provided with bench terracing and used for growing grasses, life hedges with vetiver grass and trees. The excess water can be diverted from the field to the storage pond through a water diversion bund. The stored water is used as life-saving irrigation during dry spells or to the rabi crops.

Stubble mulching retards the velocity of flowing water and increase the intake capacity of soil.

Water Harvesting and Recycling

In semi-arid areas the crop suffers due to moisture deficit at critical stages of crop growth due to erratic rainfall and long dry spells. The excess water harvested and stored in ponds can be used as life- saving irrigation. Necessary steps should be followed to check leaching or sideward seepage of water from the pond.

Inter-row water harvesting technique is very useful in dry land farming. Water stored in open furrows in a field is easily available to the crop.

Reducing Evapotranspiration

In semi-arid regions, the rate of evapotranspiration is greater than rainfall. The ET rates from soil surface can be reduced by mulching, ploughing the field from time to time, weed control and use of anti-transpirants.

Transpiration from crop surface can be reduced by using anti-transpirants materials like phenyl mercuric acetate (PMA). This waxy material cover the stomata with a thin film which are more resistant to passage of water.

Weeds compete with crops for water and nutrients. Weeds are wasteful and can be controlled by using weedicides such as attrazine.

In dry land farming mulching has several benefits:

Mulching reduces run off, increase rate of infiltration, control weeds, reduce evaporation and maintain optimum soil temperature.

Crop residues and others crop wastes such as straw, stover, saw dust and woodchips have been suggested as mulch materials.

Efficient use of water and nutrients

The crop production constraints in dry land farming are:

- Limited moisture availability
- Traditional crops and cropping system, lack of improved seeds and agronomic practices
- Poor crop stand due to soil crusting
- Low use of organic manure, in organic fertilizers and plant protection measures.

Following agronomic practices are recommended for efficient use ofwater and nutrients:

The crops and their varieties should be selected based on the water availability or length of growing season. The crops like sorghum, pear millet, groundnut, castor, arhar, safflower, mustard, cotton, etc. were found suitable for semi- arid regions.

The crops selected for this region should be: short duration, low water requirement, response to low agronomic inputs, able to escape draught, high yielding and deep rooting system.

Instead of sole cropping, intercropping of a shallow and deep rooted crop combinations like rice+arhar, maize+groundnut, sorghum+groundnut, maize+soyabean were found more remunerative in dry land areas.

The selection of a crop rotation will be based on land use capability, climate, soil type, slope, character and degree of erosion and other physical and chemical properties of soil. In a crop rotation, pulse should find a place to maintain soil fertility.

Under rainfed condition, maize-gram or jowar-gram crop rotation is feasible in deep vertisols. Rice-horse gram, maize- mustard or rice-mustard cropping system was found suitable in red soils.

Early planting of crops in rainfed areas is advantageous from the point of better crop establishment, better use of available moisture and nutrients, less incidence of pests and disease. Dry sowing of crops in red soils in anticipation of rains was found suitable. The problem of crusting can be mitigated by addition of organic manure, lime, phosphate fertilizers etc.

Dry land soils are not only thirsty, but also hungry. Balanced use of plant nutrients with organic manure is essential for higher crop productivity. Organic manures not only supply plant nutrients, but improves soil physical conditions and water holding capacity of soil. Deep placement of fertilizers in the moist zone is recommended in semi-arid tracts. Application of lime (in acid soil), secondary and micro nutrients and bio-fertilizer improved yield of several crops in dry land regions.

Management of coastal sandy soils

Coastal sandy soils occupy an area of 1.1 m ha which is about 0.3% of total geographical area of the country. It occurs along the coast line comprising sand beaches, beach ridges, sand hills, sand banks, sand bars and sand dunes. The rivers like Krishna, Cauvery, Mahanadi have sandy soils at their mouth.

Along the eastern coast, the sandy soils lie over the crystalline Archean formation known as *cuddalore* series. It is mainly ferruginous, gritty sand stone and lime stone. The original source of sandy debris is the land from where the water pass to the seashore through streams. However, the major part of sand building is derived from the sea floor by sea wave.

The wind flowing on seashore for the whole year except December and February are effective in bringing the sediments from the submarine floor near the shoreline and build up ridges, bars, barriers or sand dunes, above the sea level.

The sandy and coarse loamy coastal soils have ECe 15-25 dS m^{-1} and are characterized by chloride of Na, Mg, Ca type of salinity. Sulphate and HCO_3^- content is very small. The SAR is mostly between 25-35 and pH is around 8.0. In some areas the soils are non-saline to moderately saline, mild alkaline in reaction with SAR values less than 10. The coastal sandy soils of Goa and Kerala are mildly acidic. The Kerala tract has extensive occurrence of acid sulphate soils.

In Gopalpur coast of Odisha, sand deposits are of high industrial value. The white quartz beach sand with uniform grain and texture occurring on the Travancore coast is used for manufacture of glass. Sand containing monazite is found in the beach of Kerala and Odisha (Gopalpur). Titanium rich sand occurs on the beaches in the state of Kerala, Maharashtra, Odisha and Andhra Pradesh.

Coastal sandy soils contain heavy minerals like ilmenite, zircon, rutile, monazite, sillimanite, garnet etc. The loose sand has bulk density of 1.5 gm cm^{-3}, high infiltration rate (15 cm sec^{-1}) and very low water retention capacity. Moisture retrained at 1/3 bar is less than 5%.

Management

Deficient in plant nutrients, excessively droughty with poor moisture storage and transmission characteristics and prone to wind and water erosion, the coastal sandy soils need a careful and scientific management.

Forestry, fruit tree planting, and arable farming are some of the uses to which these soils can be put to. Forest spp. like *Prosopis juliflora*, Jhau, eucalyptus are some trees which can be taken very well. Coconut, arecanut, cashew and Jack fruits grow very well. Good Establishment of seedlings could be achieved by planting the seedling in treated pits. Pit treatment can be done with pond sediment @ 10 kg per pit or bentonite clay @ 2-3 kg/pit. This will reduce percolation loss of water, enhance soil water storage and markedly increases the growth of plants. It also reduces the loss of nutrients through leaching.

Study Questions

1. Write short notes on

i) Terracing

ii) Water harvesting

iii) Shelter belts

iv) Bio-drainage

v) Moisture deficit Index

2. Differentiate the land characteristics features between hilly areas and semi-arid regions with reference to soil and water erosion.
3. Name various land management practices recommended for (i) hilly areas (ii) water logged wet lands.
4. State different production constraints of rainfed agriculture. How will you increase the water retention capacity of soil in these areas?
5. "Dry land soils are not only thirsty, but also hungry". Comment on this statement.
6. State different ways to reduce run off loss of rainwater in rolling topography.
7. Give a brief account of industrial values of coastal sandy soils.
8. What are the problems of arid zone? Suggest some crop management practices for arid agriculture.